SCOTLAND: THE CHALLENGE OF DEVOLUTION

Scotland: the Challenge of Devolution

Edited by
ALEX WRIGHT
Department of Politics, University of Dundee

Routledge
Taylor & Francis Group

LONDON AND NEW YORK

First published 2003 by Ashgate Publishing

Reissued 2018 by Routledge
2 Park Square, Milton Park, Abingdon, Oxon OX14 4RN
711 Third Avenue, New York, NY 10017, USA

Routledge is an imprint of the Taylor & Francis Group, an informa business

Publisher's Note
The publisher has gone to great lengths to ensure the quality of this reprint but points out that some imperfections in the original copies may be apparent.

Disclaimer
The publisher has made every effort to trace copyright holders and welcomes correspondence from those they have been unable to contact.

A Library of Congress record exists under LC control number: 00134829

ISBN 13: 978-1-138-74071-6 (hbk)
ISBN 13: 978-1-138-74067-9 (pbk)
ISBN 13: 978-1-315-18334-3 (ebk)

Contents

List of Contributors

Lynn G. Bennie
Department of Politics and International Relations, University of Aberdeen.

Professor Alice Brown
Department of Politics, University of Edinburgh.

Ian Clark
Honorary Research Fellow, Department of Politics, University of Dundee.

Professor Mike Danson
Professor of Scottish and Regional Economics, University of Paisley.

Kevin Dunion
Director, Friends of the Earth Scotland.

Professor John Fairley
Professor of Public Policy and Local Economic Development, University of Strathclyde.

Karen Gilmore
Department of Industrial Relations, University of Paisley.

Professor Grant Jordan
Department of Politics and International Relations, University of Aberdeen.

Professor James G Kellas
Department of Politics, University of Glasgow.

Professor Greg Lloyd
Geddes Centre for Planning Research, School of Town and Regional Planning, University of Dundee.

Professor Arthur Midwinter
Department of Government, University of Strathclyde.

David Millar O.B.E.
Member of the Expert Panel on Procedures and Standing Orders in the Scottish Parliament.

Gerard Murray
Honorary Fellow, Unit for the Study of Government in Scotland at Edinburgh University.

Professor Peter Roberts
Geddes Centre for Planning Research, School of Town and Regional Planning, University of Dundee.

Professor Trevor Salmon
Professor of International Relations and Jean Monnet Professor, Department of Politics and International Relations, University of Aberdeen and Professor of College of Europe.

Paul Henderson Scott
Former Diplomat, author, broadcaster and former SNP spokesman on cultural affairs.

Linda Stevenson
Department of Politics and International Relations, University of Aberdeen.

Alex Wright
Department of Politics, University of Dundee.

List of Tables

List of Figures

Acknowledgements

The chapters in this book originated from a conference organised by the Department of Politics at the University of Dundee during September 1999, entitled 'The Scottish Parliament: The Consequences'. We were extremely fortunate in attracting a formidable group of scholars and public figures, virtually all of whom have contributed to this publication. We would like to thank Professor Alan Miller (Scottish Human Rights Centre) and John Berridge (University of Dundee) who gave presentations on the day but who for the best of reasons were unable to offer us a written version for publication and conversely Professor Peter Roberts who was unable to attend in September but who has contributed a chapter here. We also express our gratitude to Professor Alan Dobson and Norrie MacQueen (Department of Politics, University of Dundee), Peter Lynch (Department of Politics, University of Stirling), Charlotte Lythe (Director of the School of Contemporary European Studies, University of Dundee) and Andrew Scott (The Europa Institute, University of Edinburgh) who chaired the various panels. We are most appreciative of John McAllion, MSP MP, who, as keynote speaker set the tone for an enthusiastic and spirited day.

On a more personal level the editor is extremely grateful for the support of Susan Malloch, Helen Leslie, Cameron Ross and Richard Dunphy (Department of Politics) as well as Pam Hutton (School of Contemporary European Studies), Norma Lawson, Heather Muirhead, John and Helen Crighton, each of whom in their own way have been enormously helpful.

Since September 1999 Scotland's political landscape has changed immeasurably and this is reflected in the contributions, some of which were not completed until the following May. It therefore has the advantage of being written a while after the Parliament has been up and running. It is encouraging to see so many works being published on Scottish politics and the Parliament during the last twelve months. We expect that this volume particularly, will satisfy the interest of both the academic and non-academic reader, whether they be interested in, say, political economy, local government, international relations or current affairs.

Alex Wright, Department of Politics, University of Dundee, May 2000

List of Abbreviations

AFF	Agriculture, Fisheries and Food (Programme)
AGSD	Advisory Group on Sustainable Development
AMS	Additional Members System
AONB	Areas of Outstanding Natural Beauty
ARS	Alternative Regional Strategy
BIC	British-Irish Council
CAP	Common Agricultural Policy
CFP	Common Fisheries Policy
CoSLA	Convention of Scottish Local Authorities
CSG	Constitutional Steering Group
DETR	Department of the Environment Transport and the Regions
DOE	Department of the Environment
EAC	European Affairs Committee (of the Scottish Parliament)
EC	European Community
EMU	Economic and Monetary Union
EU	European Union
FCO	Foreign and Commonwealth Office
FIAG	Financial Issues Advisory Group
FPTP	First Past the Post
GDP	Gross Domestic Product
GERS	Government Expenditure and Revenue in Scotland
GGBR	General Government Borrowing Requirement
GOR	Government Offices for the Regions
HMG	Her Majesty's Government
HMIPI	Her Majesty's Industrial Pollution Inspectorate
ICT	Information and Communications Technologies
IPA	Involvement and Participation Association
JMC	Joint Ministerial Committee
LECs	Local Enterprise Companies
MAI	Multi-lateral Agreement on Investment
MNE	Multinational Enterprises
MoU	Memorandum of Understanding
MP	Member of Parliament
MSP	Member of the Scottish Parliament
NCG	Non-central Governments
NDPB	Non Departmental Public Body

OECD	Organisation for Economic Cooperation and Development
PM	Prime Minister
RDA	Regional Development Agency
RSPB	Royal Society for the Protection of Birds
SAC	Scottish Affairs Committee
SEA	Single European Act
SEPA	Scottish Environment Protection Agency
SGP	Scottish Green Party
SLP	Socialist Labour Party
SNA	Sub-national actor
SNH	Scottish Natural Heritage
SNP	Scottish National Party
SSA	Scottish Socialist Alliance
SSP	Scottish Socialist Party
STUC	Scottish Trades Union Congress
SWP	Socialist Workers Party
TEU	Treaty on European Union
TGWU	Transport and General Workers' Union
TUC	Trades Union Congress
UDP	Ulster Democratic Party
UK	United Kingdom
UKRep	UK Permanent Representation (in Brussels)
UUP	Ulster Unionist Party
VFM	Value-for-Money
WTO	World Trade Organisation

PART I
OVERVIEW

1 Introduction

ALEX WRIGHT

> Labour promised to deliver for Scots and it did. One of its very first priorities was a Scotland Bill giving Scotland the right to run its own affairs while still playing its full part in Britain.
> Whilst previous governments centralised power New Labour brought a radical new deal. The Scottish Parliament is just a part of a huge programme of reform bringing power back to the people
> (Building Scotland's Future, 1999, The Labour Party).

With these words 'Scottish New' Labour set about campaigning for the 1999 elections to the Scottish Parliament. They reminded Scottish voters that promises had been kept, they implied that a 'radical new deal' would be the antithesis of 'centralised power' and they portended that this was but part of a 'huge programme of reform' the object of which was to bring 'power back to the people'. Whilst much of this can be dismissed as the kind of hyperbole that exists in any party election manifesto, it did reflect the tenor of Labour's aspirations and they were written at a moment in our history when a page had been turned. More particularly they caught the mood of the time - no longer would Scotland be governed in a way that was alien to the values and wishes of the Scottish people. As Gordon Brown and Douglas Alexander intimated in *New Scotland New Britain*,

> The constitutional consequences of Mrs Thatcher - unintended by her though they were - are that in Scotland at least, another Mrs Thatcher can never again represent the same threat. No second Mrs Thatcher could ever inflict such damage on Scottish civic life again (Brown and Alexander 1999 p.10).

There is certainly a grain of truth in this, but ironically not necessarily in the way that the authors intended. Now that there is a Scottish Parliament we have an elected body which will ensure that public policy in Scotland is more democratically 'open' (Paterson 1998 p.54) and in so doing if necessary it could act as the rallying point for Scottish grievance if ever again if a UK prime minister overstepped the mark north of the Border.

But no sooner had new Labour's programme of constitutional reform got under way, than there was growing scepticism concerning its promise to redistribute power away from 'London' or more narrowly from the prime minister himself. In Wales Rhodri Morgan the popular choice for leader of the Labour administration was denied the chance when 'Millbank' interfered with the electoral arrangements, thereby ensuring that its own placeman Alun Michael was selected. A sizeable body of the Welsh electorate, already uncertain about the need for a Welsh assembly responded by voting for Plaid Cymru and Alun Michael himself was eventually forced to resign in February 2000. The selection of London Mayor has followed a similar trajectory with Ken Livingstone viewed as 'off message' by Downing Street with the result that Frank Dobson became the Labour candidate. Once again, however, events slid out of Blair's control when Livingstone resigned and declared that he would be an independent candidate, thereby splitting Labours vote. There are two conclusions to be drawn from this. First Tony Blair is naively attempting to retain executive influence despite creating new institutions of devolved government - this is inherently contradictory. Second he has misunderstood the political consequences of constitutional reform - something that has not been lost on electorates and politicians at the territorial level. At the time of writing the House of Lords reform has run into the sand with critics complaining that prime ministerial appointees have replaced the hereditary peers. Only Scotland, it would appear has been free from prime ministerial interference but even here Labour's 'programme' has not been without its set-backs.

Whilst Donald Dewar appeared to have straddled the divide between 'old' and 'new' Labour and was seen to be a 'safe pair of hands', in the run up to the elections in May the party itself was disunited (Taylor 1999 p.152-3) and afterwards it appeared increasingly accident prone. If the resignation of Alex Rowley as General Secretary of the Scottish party just weeks after the 1999 election raised eyebrows, the dismissal of John Rafferty, one of Mr Dewar's closest aides, sparked public debate. By the first few months of 2000 the credibility of the First Minister himself was on the line as the Executive faced vituperative attacks from the tabloids and from populists like Brian Souter over its commitment to repeal Section 28 (which prevented schools from promoting homosexuality in the curriculum). Regardless of whether the original legislation was flawed, the Executive appeared inept and was wrong-footed over its handling of the fall-out that followed. Why was it, some asked, that the Executive should find itself embroiled in such a cause so early on in the Parliament at a time when much needed to be done if devolution was to prove its worth with the electorate? That was subsequently overshadowed by the escalating costs (c£200m) of the construction of the Parliament and its ancillary buildings. All-in-all it seems that the Parliament had yet to begin to meet the (longer-term) challenge posed by James Mitchell, namely that it had to 'cut out a role for itself, some autonomy, without having to spend considerable sums of money' (Mitchell J 1998 p.76).

Of all the aspirations that greeted the new Parliament, foremost was the desire that it would not be a clone of Westminster. There were a variety of aspects to this. It was expected that Holyrood would be less confrontational and this was reflected in the semi-circular design of the Parliament's debating chamber at its temporary home on the Mound. This contrasts with Westminster where Government and Opposition sit opposite each other but it does not follow that debate will be any less lively. Despite the civility of proceedings in the debating chamber the SNP has shown itself to be an effective and mature opposition that has enthusiastically harried the Liberal Democrat - Labour coalition which forms the Scottish Executive. There was also the hope that there would be a different style of politics at Holyrood - less sleaze and greater inclusivity. Underpinning this was the notion that those that now governed Scotland would recognise that circumstances had changed. Yet, as David Millar observes in Chapter 2, maybe we have been too optimistic.

Twelve months after the May elections in 1999 the electorate was not slow to register its disappointment with Labour. The Party lost the Ayr by-election in the spring of 2000 - albeit that as a 'marginal' it was a vulnerable seat - and by April the SNP had overtaken Labour in the polls for the Scottish Parliament (The Herald 03/04/00). Mid-term blues maybe, as Bernard Crick observed. But he also warned that this was an indication of a sharp loss of confidence in the authors of constitutional change as a result of an incomprehension of devolution by Millbank and its misguided mission to treat its core supporters across the whole of the UK as 'Middle England' (The Herald 05/04/00). Belatedly, the Prime Minister conceded "Essentially you have to let go of it with devolution" (The Times 11/04/00).

'Britain' remains integral to Labour's modernisation plans, as Tony Blair has been at pains to emphasise recently, but Scots understandably identify more with Holyrood than ever was the case with Westminster. This cannot be merely attributed to its physical proximity but the fact that it is the collective and legitimate democratic voice of Scotland - something that has been absent until now. So what will become of Scottish allegiance to Britain? Gordon Brown, the British Chancellor has argued that 'Britishness' is not so much concerned with institutions but shared values and historical experience between the component parts of the Union. That being so and drawing on the work of Johnathan Sacks in the *Politics of Hope* he then observed that the 'British' were moving from a 'contractual' political settlement based on political institutions which have been established out of self interest to a 'covenant society' based on shared values and a common purpose (The Times, 10/01/00). Although there are a great many similarities, Scotland *is* different compared to England - albeit to a lesser extent in terms of its values than by virtue of the national dimension to its political culture (Brown *et al* 1999) - added to which there are considerable variations within Scotland itself. Does Tony Blair's Government really possess the vision and political nous to appreciate that Scottish politics is distinct from the rest of the UK and that this led to devolution?

Despite Blair's promise to stop 'meddling', the impression persists that Labour has misunderstood or chosen to ignore the consequences of devolution. Brown and Alexander viewed constitutional reform in terms of 'a new relationship in which the individual is enhanced by membership of their community, and the state enables and empowers rather than controls or directs' (Brown and Alexander 1999 p.36). But 'enable' and 'empower' are little more than political jargon that disregards the discrete functions of the various branches of government - something that John Fairley and Greg Lloyd examine, respectively, in Chapters 7 and 8. Scotland now has an additional tier of democratically elected government and this will have all sorts of repercussions not just for citizens but also for the various institutions that were there before. What will the effect be on local government, for instance? Will some of its powers be re-assigned to the Scottish Executive and if so will local government be the worse for it? Quangos long demonised by opposition parties during the era of the Conservative 'Raj' look vulnerable on the surface because they too might be subsumed by the Executive. More latently there may be good reason for retaining them if it enables the Executive to circumvent non-aligned local councils. Transcending that though is the *real politic* that relations between Scotland and the UK have changed irredeemably and so too has the character of Scottish politics - the elite in London will be more constrained north of the Border than has been the case hitherto.

Yet Brian Taylor alluded that the London wing of the Labour party could 'play down the distinctive nature of Scottish politics' and they viewed their counterparts in Scotland as 'perverse in stressing the Scottish political dimension' (Taylor 1999 p.151). If that is so why was it so keen on devolution? In practice Labour had little option but to press ahead with a Scottish Parliament. Electorally it would have been folly not to do so, given the strength of opinion in Scotland coupled with the value of Scottish seats to Labour - and - there was also the legacy of John Smith the former leader who was committed to devolution - which Blair inherited.

Even so, there was a strand of opinion within the Labour party which was opposed to the 'Balkanisation of Britain' (Taylor 1999, p.149). So constitutional change was subsumed in new Labour's ideology of

'Modernisation'- this was not intended to pander to nationalism it was to be 'regenerative'. Modernisation of the Labour party itself went hand in hand with modernisation of the British state but revealingly, Blair's closest advisors apparently devoted little thought to Scotland or Wales. Phillip Gould, a member of Blair's inner sanctum made virtually no mention of Scotland whatsoever in his book *The Unfinished Revolution* (save the debacle over the Parliament being compared [erroneously] to a parish council - 'probably the least successful day' of the 1997 UK election campaign (Gould 1998 p. 361)) nor was there much, if any, mention of Wales. Devolution was first and foremost an idea rather than an ideal for 'London' Labour, inasmuch as Gould explained: 'Tony Blair was obsessed with winning the battle of ideas ..... Ideas which have the power to dominate the political agenda' (Gould 1998 p.231). Constitutional reform, therefore was a means to an end - an election victory for Labour - and in so doing it would act as the saviour of a British state that could no longer survive as a unitarist polity. As Jack McConnell a minister in the Scottish Executive later observed, 'In the new century, modernisers must renew and rebuild to make the new parliament succeed and stabilise the UK' (McConnell 1999, p.68).

Devolution it can be said therefore is as much about 'stabilising' the UK as 'empowering' Scots, or the Welsh for that matter. Whilst Brown and Alexander viewed it as 'no half way house' to separatism (Brown and Alexander 1999 p.46), James Kellas who writes in Chapter 3 of this book has his doubts. He is not the only one. James Mitchell noted, not only is the Scottish Parliament on sufferance but so too is the British State itself. 'If the perceived failure of Britain lies behind the demand for Scottish constitutional change, then it might be expected that perceived continued failure will result in increased demands' (Mitchell 1998, p.81). Tom Nairn, who for some time has predicted the demise of the British state, argued that the Scottish Parliament co-existed with a 'faltering multi-national state' where 'constitutional renovation [has] been loudly spoken of, but then half abandoned in disarray leaving all the most crucial modernisation-problems in limbo. No formula for 'stability' is visible here (other than robotic conformity to the will power of Blair's 'project')' (Nairn 2000 p.256). That may be true but devolution is the product of a 'negotiated process' involving compromise and bargaining that has long been the hallmark of the relationship between the UK and Scotland (Paterson 1998 p.54). If devolution fails to meet the aspirations of Scots when so much has been promised by the UK and Scottish Governments then the existing constitutional arrangement would be unfinished business.

Labour's open-ended programme has engendered its own terminology. Some commentators refer to it as 'asymmetrical government' (e.g. England is yet to enjoy devolution) and they argue that not only does it have historical antecedents but that it may be the most 'practical' solution to variable demands for territorial autonomy within a single state (Keating 1998 p.213). Others, whilst accepting the asymmetrical connotation, prefer the term quasi-federalism by virtue of the pseudo federal characteristics of the UK following constitutional reform (Hazell 1999 p.231). Devolution is quite distinct from Federalism, however. Vernon Bogdanor defined it thus:

> Devolution involves the creation of an elected body, subordinate to [Westminster] Parliament. It therefore seeks to preserve intact that central feature of the British Constitution, the supremacy of Parliament. Devolution is to be distinguished from federalism, which would divide, not devolve, supreme power between Westminster and various regional or provincial parliaments   (Bogdanor 1999 p.3).

Constitutionally the Scottish Parliament may be subordinate to Westminster but the political landscape has changed irredeemably. Tom Nairn rightly claimed 'for the UK state power devolved is emphatically *not* power retained' (Nairn 2000 p.109). That is to say than even if the current political elite in Millbank and Downing Street have failed to appreciate it, as suggested above, there really has been a transfer of power from London to Scotland. The consequences of this are only just beginning to emerge as the Scottish Parliament finds its feet.

In the meantime the UK finds itself in the midst of a constitutional 'anomaly'. As Neil McCormick warned, will the 'pragmatism' that has long been the *leitmotiv* of England's approach to constitutional affairs survive 'the process of switching from inconspicuous anomaly to highly visible anomaly in the constitutional position of the elements in the union'? (McCormick 1999 p.195). How, if at all, will England respond and what will the consequences be for Scotland? Peter Roberts suggests in the final chapter of this book that Scotland has been the pacesetter for the other territories of the UK and if they were to catch up, this might fuel demand for greater autonomy north of the Border. There has already been talk of an English Grand Committee, which rather ignores the relative impotence of its Scottish counterpart. The West Lothian Question - which relates to the right of Scottish MPs to vote on purely English matters when English MPs no longer have the authority to do the same for those areas of competence that have been devolved to Scotland - has now been re-dubbed the 'English Question'. This is something of a paradox when the votes of English MPs had ensured that Westminster's policies had been imposed on Scotland during the 1980s and 1990s even after the Scottish electorate had rejected 'Thatcherism' at successive elections. But the issue is unlikely to go away until something is done about it and if nothing is done then at some point in time there is the threat of a constitutional crisis.

Just as controversial is the issue of the bloc grant and public expenditure both of which are inherently connected to the 'English Question'. For some time the Tories and Labour MPs in the North of England have called into question how the constituent parts of the UK will be funded by central government. They are not alone: Ken Livingstone was reported to have called for change during his campaign to be Mayor of London (The Herald 31/03/00). The politics of devolution finance is something that will become increasingly contentious no matter who is in government in London and Holyrood, not least because the Treasury will have its own agenda. Devolution finance will be analysed in detail by Arthur Midwinter in Chapter 16. With Scotland's Parliament in *situ* there have been growing demands for parity - especially from the less prosperous parts of the UK in the North of England. The North East especially has watched the evolution of devolution with a mixture of alarm and interest. In the absence of its own parliament how can it compete for funding and political influence compared to Scotland or Wales?

That said, the efficacy of the Scottish Parliament itself is debatable. Notwithstanding the fact that a single democratically elected body has the legitimate right to speak for Scotland on any subject it so chooses there has been criticism over Holyrood's powers. Although considerable groundwork had been undertaken by the Constitutional Convention before Labour won the UK election in 1997, critics complain that too little thought was given to the powers of the Parliament, and this was especially apparent in relation to its ability to vary tax (Mitchell 1999 p.30). Can Holyrood really be referred to as a parliament when its capacity to raise revenue is less than that of a local authority? This will have ramifications for the future and already there has been some debate as to what will happen when the tax varying powers are first used. For instance it is likely to mean that if two individuals work for the same company north of the Border - in Hawick for example - but one resides in Scotland and the other commutes from England, that their net pay will differ. But that is only the tip of the iceberg as to how taxation matters to Scotland; devolution and the political economy of Scotland, is analysed by Mike Danson and Karen Gilmore in Chapter 15.

It is easy to fall into the trap of believing that devolution simply relates to Scotland and the UK. This impression is reinforced by the fact that Foreign Relations has been reserved to Westminster. That is to say that the Foreign and Commonwealth Office is still responsible for the conduct of foreign policy and by default relations with the European Union (EU). In practice this is not so, in part because Scotland now has a parliament and by virtue of other factors such as 'globalisation' which 'draw regions into the international arena' (Keating 1999 p.14). For example international trade disputes can have an enormous impact on Scotland and this will be of concern to both the Scottish Executive and MSPs - as Trevor Salmon observes in Chapter 11. Even though the 'regionalisation' of the EU may have stalled in recent years and Scotland's influence in the EU is questionable despite devolution, as Alex Wright suggests in Chapter 10, the Scottish Executive and Members of the Scottish Parliament (MSP) will inevitably be engaged in influencing EU policy. The resumption of direct rule by Westminster (at the time of writing) is indicative of the difficulties that have beset devolution in Northern Ireland and has ramifications for the British Irish Council, of which Scotland is a member. But according to Gerard Murray in Chapter 9, that Scotland now has its own parliament will transform the political relationship between Scotland and Ireland. So once again devolution will have international repercussions and must not be seen purely as domestic politics.

For so long denied a parliament of its own Scotland now has the opportunity to govern itself. Yet there is the technical professional issue regarding the extent to which a Scottish Parliament will be able to construct a series of policies that reflect the real problems and opportunities evident in Scotland, other than having to work with the 'one-size-fits-all' set of policies that are UK-wide. To an extent Scotland has made progress on this front prior to devolution. Will the situation improve now? Equally, will 'devolution' really bring government any 'closer' to the Scottish people given that Scotland has had its own territorial office in Edinburgh for the last 100 years or so and is devolution tantamount to self-government? Self-government is not merely concerned with a territorial political elite securing the right to manage the affairs of a country like Scotland, rather it relates to the extent to which citizens can govern themselves via their democratic institutions (Mitchell 1996). By default this implies that a participatory form of democracy relates more to self-government than a representative one in as much as citizens themselves should be engaged more directly in the processes of government.

By having a parliament in Edinburgh as opposed to Westminster it is possible for more Scots to involve themselves physically in their governance than was the case hitherto. All sorts of ideas have been advanced with a view to ensuring that Scotland has a '*citizens*' Parliament rather than one this is primarily for politicians. It was hoped that citizens' juries would be established in order that parliamentarians could sound out sections of the electorate. The underlying objective was that MSPs should not become too remote from the Scottish society. Scotland will shortly have a Civic Forum - an intention being that civic society can interface with the

Parliament without the mediation of individual MSPs. This is supposed to be part and parcel of the new Scottish politics but whether it spells the demise of pressure groups is questionable. This and the potential for tension between participatory and representative democracy is analysed in detail by Grant Jordan and Linda Stevenson in Chapter 12.

In the meantime the most visible development, is the composition of the Parliament and Executive thanks to the inception of a proportional voting system which led to a coalition government. Ian Clark examines the outcome of the 1999 election in Chapter 5 and Lynn Bennie assesses the ramifications for the 'small parties' in Scotland in the following chapter. An intriguing question is whether MSPs will replicate their colleagues at Westminster and be hidebound by the party system or whether they will be influential players in their own right. Given that many of them have no parliamentary experience at all there is cause for concern but there are grounds for optimism too. That Scotland is governed by a coalition means that the Parliament cannot be dominated by a single party and as Brian Taylor has affirmed, MSPs see themselves as 'players' rather than 'observers' (Taylor B. 1999 p. 262). This was borne out in April 2000 when the Executive backed down in the face of parliamentary opposition to its attempt at impeding Tommy Sheridan's bill abolishing warrant sales. Others are more cautious. For example Kevin Dunion writing in Chapter 13 on environmental policy and sustainability argues that there needs to be a sense of enthusiasm at Holyrood if issues like this are to be addressed adequately, even though competence may reside elsewhere.

Will devolution really make any difference and if so how? On the basis that most power in Scotland rests with the ministers and civil servants in the Executive, rather than with parliamentarians this may be questionable. To some extent 'legislative' devolution stands to make little difference to the governance of Scotland when the country enjoyed a fair measure of administrative autonomy anyway prior to 1999 (Paterson L. 1994). Then there is London's apparent resistance to and misconception of devolution. Yet even Paul Scott, who is critical of the Parliament's powers, acknowledges that its existence 'changes everything' when he writes about Scottish culture and the Scottish Parliament in Chapter 14. In a similar vein Alice Brown suggests in Chapter 4 that 'Scottish Politics can never be the same again'. It is difficult to be other than optimistic; despite the difficulties of recent months, already the Parliament is well on the way to establishing its credibility.

This book is a preliminary analysis of the first year of the Scottish Parliament and the challenges that confront it. That said, the contributing authors, each of whom are specialists in their field, offer the reader an informed view as to what the future holds for the Parliament and the potential ramifications of devolution.

Note

The Author thanks John Fairley and Peter Roberts for their advice when this chapter was first drafted.

References

Aldecoa, F. and Keating, M. (1999), *Paradiplomacy in Action. The Foreign Relations of Subnational Governments*, Frank Cass, London.

Bogdanor, V. (1999), *Devolution in the United Kingdom*, Oxford University Press, Oxford.

Brown, A. et al (1999), *The Scottish Electorate*, Macmillan Press, Basingstoke.

Brown, G. and Alexander, D. (1999), *New Scotland. New Britain*, The Smith Institute, London.

Building Scotland's Future (1999), Scottish New Labour.

Elcock, H. and Keating, M (eds) (1998), *Remaking the Union. Devolution and British Politics in the 1990s*, Frank Cass, London.

Gould, P. (1998), *The Unfinished Revolution. How the Modernisers Saved the Labour Party*, Abacus, London.

Hassan, G. and Warhurst, C. (eds) (1999), *A Modernisers Guide to Scotland. A Different Future*, Centre for Scottish Public Policy and the Big Issue, Edinburgh.

Hazell, R. (ed) (1999), *Constitutional Futures*, Oxford University Press, Oxford.

The Herald, March 31 2000, Livingstone says London pays for Scottish Health, page 6.

The Herald, April 3 2000, Malcom Dickson, The Honeymoon is over as Labour records lowest share of vote, page 6.

The Herald, April 5 2000, Bernard Crick, Letters to the Editor, page 16.

Keating, M. (1998), 'What's Wrong with Asymmetrical Government', in H. Elcock and M. Keating (eds), *Remaking the Union. Devolution and British Politics in the 1990s*, Frank Cass, London.

Keating, M. (1999), 'Regions and International Affairs: Motives, Opportunities and Strategies', in F. Aldecoa and M. Keating, *Paradiplomacy in Action. The Foreign Relations of Subnational Governments*, Frank Cass, London.

MacCormick, N. (1999), *Questioning Sovereignty*, Oxford University Press, Oxford.

McConnell, J. (1999), 'Modernising the Modernisers', in G. Hassan and C. Warhurst (eds), *A Modernisers Guide to Scotland. A Different Future*, Centre for Scottish Public Policy and the Big Issue, Edinburgh.

Mitchell, J. (1996), *Strategies for Self-Government*, Polygon, Edinburgh.

Mitchell, J. (1998), 'What could a Scottish Parliament do?' in H. Elcock and M. Keating (eds) *Remaking the Union. Devolution and British Politics in the 1990s*, Frank Cass, London.

Mitchell, J. (1999), 'Consensus: What Consensus?' in G. Hassan and C. Warhurst (eds), *A Modernisers Guide to Scotland. A Different Future*, Centre for Scottish Public Policy and the Big Issue, Edinburgh.

Nairn, T. (2000), *After Britain. New Labour and the Return of Scotland*, Granta Books, London.

Paterson, L. (1994), *The Autonomy of Modern Scotland*, Edinburgh University Press, Edinburgh.

Paterson, L. (1998), 'Scottish Home Rule: a Radical Break or Pragmatic Adjustment?' in H. Elcock and M. Keating (eds), *Remaking the Union. Devolution and British Politics in the 1990s*, Frank Cass, London.

Taylor, B. (1999), *The Scottish Parliament*, Polygon, Edinburgh.

The Times, January 10 2000, Gordon Brown, This is the time to start building a Greater Britain, page 15.

The Times April 11 2000, Tom Baldwin and Roland Watson, Blair pledges to stop devolution meddling, page15.

2 Scotland's Parliament: A Mini Westminster, or a Model for Democracy?

DAVID MILLAR

Introduction

For the purpose of this chapter, it is assumed that the reader is sufficiently acquainted with the history and practices of the House of Lords and the House of Commons to obviate the need to describe them. The intention of the present government to reform the House of Lords has been proclaimed, but the report of the Royal Commission thereupon is currently awaited, and thereafter a Joint Committee of both Houses will pronounce upon it. We may, probably with safety, assume that the powers of the Lords will not suffer great change.

But what attributes should the Scottish Parliament possess to earn the sobriquet of a 'model for democracy'? This chapter argues that the four key principles set out in the report of the Constitutional Steering Group (CSG) constitute the paradigm for such a model, were the Scottish Parliament to succeed in putting them into practice effectively (Scottish Office, 1998a). These principles are acceptable as criteria as they are in turn based on the work of the Scottish Constitutional Convention from 1990 - 1995 (Scottish Constitutional Convention, 1995).

The Convention achieved extraordinary success in bringing together representatives of the great majority of the Scottish people both from the House of Commons and from Scottish local authorities. The structure of Scottish society was represented by representatives of the Churches, the trades unions, the political parties, ethnic minorities, small businesses, legal bodies, the voluntary sector and others. Reaching decisions by consensus over a period of six years, the Convention hammered out agreements on the size, character and style of the Parliament, on the electoral system to be used, on its powers and functions, on the application of equal opportunities, on the principal procedures, finance and several other aspects of its operation.

These agreements were encapsulated in the summer of 1997 in a government White Paper (Scottish Office 1997), on which basis a referendum in September sought the opinions of the Scottish people on the creation of the Parliament and on its tax-varying powers. Thus the concepts underlying the four key principles enunciated in the CSG report evolved over a long period of debate, were brought before the Scottish people in three major publications and were examined at innumerable conferences, committee meetings and discussions over the length and breadth of the country, culminating in an intensive consultation exercise by the Consultation Group itself.

The Four Key Principles

The four key principles are:

> The Scottish Parliament should embody and reflect the sharing of power between the people of Scotland, the legislators and the Scottish Executive;

> The Scottish Executive should be accountable to the Scottish Parliament, and the Parliament and Executive should be accountable to the people of Scotland;

> The Scottish Parliament should be accessible, open, responsive and develop procedures which make possible a participative approach to the development. consideration and scrutiny of policy and legislation;

> The Scottish Parliament in its operation and its appointments should recognise the need to promote equal opportunities for all.

It would be too facile at this stage simply to dismiss, having considered these principles, any possibility of the Parliament becoming a 'mini-Westminster', even though the CSG elaborated a procedural structure for the Parliament based upon them. The sense of the Constitutional Convention's approach was to argue that, given the 300-year period during which the Scottish Parliament was 'adjourned', the Parliament of 1999 could start with - in Secretary of State Dewar's words in December 1997 - 'a clean sheet'. Given this situation, the CSG was obliged, by public opinion and by necessity, to take a stand by shaping a Parliament as distinct from Westminster, and as modern and innovative as possible. By thus doing, the danger of creating a 'mini-Westminster' was greatly lessened - but perhaps not entirely dispelled, as may become evident.

One of the motivating forces behind the drive for a totally fresh start, based on the most open and innovative Parliamentary procedures, was the awareness of the two political parties brought together in the Constitutional Convention that a democratic and devolutionary regime for Scotland must work, and be seen to be working. Neither the Scottish National Party nor the Scottish Conservative and Unionist Party had played any part in the Convention. The two 'parties of devolution' - the Scottish Labour Party and the Scottish Liberal Democrats - had therefore not only to outflank them in making devolution work, but had to convince their own sister parties in the UK, and the Westminster Parliament, on several counts. These were so that the Scottish people's leaders could elaborate working procedures for a Parliament based on wide consultation, which would, by operating effectively, withstand political assault, media cynicism and latent scepticism among the Scottish electorate.

While it is of course premature to attempt a judgement on whether this bold experiment will be successful. The latter part of this chapter will discuss Parliament's early teething troubles and assess their significance for the future. In effect a totally new Parliament should arguably be allowed five years to find its feet before any substantive assessment of its efficacy is undertaken.

Implementation of the Key Principles

The Consultative Steering Group expressed its objective in a simple sentence:

> Our aim has been to try to capture, in the nuts and bolts of Parliamentary procedure, some of the high aspirations for a better, more responsive and more truly democratic system of government that have informed the movement for constitutional change in Scotland (Scottish Office, 1998b).

Sharing Power

Principle one relates to the sharing of power. The first point to make - and it is often overlooked - is that a system of proportional representation, by relating seats won to votes cast, is the best guarantee of the closest possible affinity between the will of the electors and the government elected to power. At the election on 6 May 1999, the Scottish Labour Party won 38.8% of the vote for constituency seats and the Scottish Liberal Democrats 14.2% of this vote. Thus the partnership in government between these parties is sustained by 53% of the votes cast for constituency seats, expressed in 73 seats out of the 129 in the Parliament (including seats won on regional lists). This is a major advance on the mandates won by a governing party in the House of Commons, only one of which since 1945 has gained 50% of the vote; in 1997 Labour won only 44%, for example.

Thus from the start, power is shared more equably between the people, the legislators and the Scottish Executive.

The second manifestation of power sharing is the creation of a Parliamentary Bureau, or Business Committee, whose proposals on the priorities and timing of business on the Floor and in committees are subject to a vote of Parliament. The task of the Bureau is to balance the need for the Executive to govern, against the demands of individual Members for Parliamentary time, and those of Parliament's committees.

The Presiding Officer chairs the Bureau and has the duty of acting impartially to protect the rights of all MSPs. He has considerably greater powers and influence than the Speaker of the House of Commons, and is a major player in assuring the sharing of power.

The CSG, acting on the Convention's proposals, insisted that the Parliament should - like all Continental Parliaments - be committee-based. All-purpose committees, acting through open and responsive procedures, are multi-functional and ought in time to enable their members to develop expertise in their subject area sufficient to enable them rigorously to scrutinise and criticise the Executive's policies and administration. Again, power will be shared.

As the CSG report states, power-sharing is 'also about the empowerment of external groups and individuals in all sectors of Scottish society' (Scottish Office, 1998). The proposal for a Civic Forum, bringing together groups and individuals in Scottish civic society in order to engage with the Parliament, has come to fruition and has won preliminary funding. The Forum should facilitate consultation on policy, legislative proposals and implementation of legislation.

Finally, power is shared with groups and individuals through public petitions to Parliament, on which the Public Petitions Committee - unlike that in the House of Commons - is empowered to take any action it considers necessary to bring the complaint in the petition to the attention of Parliament or the Executive.

Accountability: of Executive to Parliament

Principle two lays down the principle of accountability, of Executive to Parliament, and of both to the people. A Code of Conduct for MSPs is to be matched by the strong committees, by Parliamentary Questions, audit procedures and by a strong European committee dedicated to examination of European issues.

The detailed manner in which the CSG's Report defines accountability brings more than one echo of the procedure of the European Parliament. It has since the mid-1980s obliged the Commission to present an annual legislative programme. A similar provision is contained in Rule 5.7 of the Standing Orders, which also provides for a debate. Similarly Rule 5.8 sets out the arrangements for a three-stage consideration of financial business. In May and June each year outline proposals for expenditure in the coming year are considered; in October and November draft budgets of public expenditure are considered by committees; and in January and February the main Budget bill is taken in committee and on the Floor.

These Rules show the Executive submitting detailed financial proposals to Parliament in accordance with a timetable, which should enable the committees to hold the Executive to account in a way which the House of Commons has not done for at least 30 years.

General debates and ministerial statements give parties and back-benchers the opportunity to question the Executive in a normal Parliamentary way and are provided for by Rules 8.1 to 8.7 and 13.2. Motions of no confidence are designed to call the actions of the Executive or of an individual minister to account in the most direct fashion possible. A motion of no confidence supported by 25 members (i.e. about one fifth of the total) is automatically included in the business programme (Rule 8.12).

This represents an advance on the Commons procedure, where only a censure motion on the government moved by the leader of the Opposition is given time for debate within so-called 'government time'. Thus accountability of the Scottish Executive to the Parliament is sharper than is provided for by the equivalent procedures in the Commons.

These are good examples of accountability of Executive to Parliament in Scotland, and there are many others. Another broad swathe of procedures enables the all-purpose committees of Parliament to keep the Executive accountable to them. When under Rule 6.2.2 b), c) or d) a committee is considering proposals for legislation before the Scottish or the UK Parliament, or European Communities proposals or policies, or reform of the law, a minister or junior minister is entitled to speak in the committee, but not vote (Rule 12.2.3). This Rule governs the relationship between the committees and the Executive which, as we shall see later, became a bone of contention in the autumn of 1999. The Rules neither require ministers to attend committees, nor endow the committees with power to insist that ministers must comply with invitations to attend. Thus the concept of accountability enshrined in key principle two comes into play when the Rules - correctly - leave it to ministers and committees to establish mutual working relationships.

The right of members to put questions to ministers is more closely regulated by the Rules, which differentiate between Question Time and Open Question Time. At the former, supplementary questions may only be put by the original questioner. At the latter, any member may put a supplementary question. Urgent oral questions may be tabled for answer the same day; and the Presiding Officer may be questioned on a matter concerning the Parliamentary corporation (the body responsible for administering Parliament) or the staff of Parliament. Thus accountability to the Parliament extends to the Presiding Officer as well as to the Executive.

On Wednesday and Thursday, the two plenary sitting days, the last half hour of business comprises a motion on a constituency or local matter moved by a back bench member. This procedure seems to be working well, as on several occasions between May and November 1999 up to five speakers after the mover of the motion made brief interventions before the minister replied.

One of the thorniest of problems of accountability concerns European Community (EC) or European Union (EU) matters. The European Committee set up under Rule 6.8 comprises members already appointed to other committees, in order to ensure that the policy areas within the latters' remit are represented within the European Committee. The Committee's task of holding the Executive to account on European matters is formidable, by reason of the number of documents to be examined, the necessity of collaborating with other committees and the fact that, at the end of the day, responsibility for relations with the EU rests with the UK government, whose actions can only be monitored effectively by the UK Parliament, not by the Scottish Parliament. Thus the European Committee will not only have to question the Scottish Executive, but establish working relations with the European Committees of both Commons and Lords and with the institutions of the EU.

Accountability: of Parliament to the People

Parliament is fundamentally accountable to the electors at elections held every four years. But this is to fall into Westminster-oriented thinking: the Scottish Parliament must - to be acceptable to the people - be actively accountable. The electoral system produces 73 members elected by the first-past-the-post system in the Westminster constituencies, and they are primarily accountable to their constituents. Members elected by regional list appear to regard themselves as being accountable to electors in their region, although the large areas of some regions, such as Highlands and Islands, and South of Scotland, militate against direct accountability to any but a section of the electors.

As an institution, Parliament, under key principle two, has a duty to pursue accountability however. This it can do largely by use of the procedures set out in earlier paragraphs of this section. That is to say that, by holding the Executive to account, Parliament acts on behalf of the people.

But Parliament must also account for its own administration and finance to the people. Such matters as the salaries and allowances of members, the accommodation and equipment provided for their official use and the number and cost of Parliamentary staff and of members' personal staff are of great concern to taxpayers. Likewise the expense of the new Parliamentary building and the costs of administering Parliament are matters which Parliament must explain and justify to the electors. There was lively controversy in June 1999 when members debated and voted on their salaries and allowances and on the site of the new Parliamentary building.

Access, Openness and Information

Perhaps the most challenging principle to implement will be the third, which insists that Parliament must be accessible, open and responsive and develop participative procedures regarding the formulation, discussion and scrutiny of both policy and legislation. But, as the CSG points out,

> It is clear when we consider the responsibilities which lie with the Scottish Executive [e.g. for implementing policies and developing legislation], that the way it operates will have considerable influence on the way the Scottish Parliament is perceived.

Thus the Executive must practise openness and be accessible: if it does not, the Parliament's right under Section 23 of the Scotland Act 1998 is to oblige it so to function: but if Parliament does not succeed in this, it will itself be blamed. This is a vicious circle, in which the House of Commons has been imprisoned for years, and from which the Scottish Parliament is already suffering, as we shall see.

The third principle can be realised by early, wide and timeous consultation, by both Executive and Parliament's committees, with groups and individuals who have an opinion to offer. Much can be done by information and communications technologies. But if only about 25% of Scots own a computer, traditional methods of circulating information must continue to be used.

Civic education in schools was emphasised by many of those consulted in the summer of 1998 by the CSG as a means of encouraging participation in the work of Parliament.

The Scottish Constitutional Convention always emphasised the importance of accessibility of committees meeting outside Edinburgh and seeking the views of citizens in different parts of Scotland. Another means of assuring accessibility is for Parliament and its committees to meet during normal business hours, with proceedings normally in public. Procedures and practices should be simple and uncluttered by pedantic ritual: Members should refer to each other by name.

It is not too much to say, on the threshold of the 21st century, that it is widely regarded as the duty of Parliaments and their dependent bodies, such as committees, to meet in public except in exceptional circumstances.

Much thought was devoted to creating a Parliamentary Information Service endowed with the most up-to-date information and communications technologies (ICT). The Consultative Steering Group wished such technologies to support the business of Parliament and MSPs and to support on-line access by the public to information about Parliamentary business and about MSPs. ICT should, in the CSG's view, also facilitate communications with the UK Parliament, the Assemblies in Wales and Northern Ireland, EU institutions, local authorities, government departments and organisations and, not least, with the public, businesses and the voluntary sector.

Budgetary and time constraints apply to this area - perhaps particularly owing to the complexity of installing and launching ICT. But good progress has been made along many of the avenues defined by the CSG (Steel).

Obviously the full implementation of these avenues of access and openness must await the completion of the new Parliament building at Holyrood. Meanwhile, the media, like the whole Parliament, is having to contend with temporary accommodation and transitional facilities.

Equal Opportunities

The fourth key principle called on Parliament to recognise the need to promote equal opportunities for all. Rule 6.9 provides for the establishment of an Equal Opportunities Committee, with a remit to report on equal opportunities, and their observance within the Parliament. The principal manifestations of a striving to promote equal opportunities have been the decisions by Parliament, based on the CSG Report, not to sit during school holidays, and to meet between the hours of 9.30 a.m. and 5.30 p.m. from Tuesdays to Thursdays, and on Monday afternoons and Friday mornings. These hours were chosen to permit as many members as possible to return home each evening, and to permit those from further parts of the country to travel to and from Parliament at reasonable hours.

The Principles in Practice: Towards Democracy

It is of course premature to try to make a judgement about the degree to which Parliament is living up to the high expectations which surround it.

After five months of work (at the time of writing), only straws in the wind can be discerned. But from the preceding sections, it might be supposed that the Parliament is already acting as a 'model for democracy' and has so far eschewed practices which would indicate a falling back to being a 'mini-Westminster'. Regrettably that supposition would not be entirely accurate.

As regards the first key principle on sharing power the indications are positive. The Presiding Officer and Deputy Presiding Officers have set excellent examples in both leading Parliament and speaking on its behalf, and guiding members through the complexities of procedures completely novel to them, and indeed novel to any elected body in the United Kingdom.

The Presiding Officer has spoken forcefully on behalf of Parliament, for example in attacking media reporting of the debates on members' salaries and allowances, describing one tabloid newspaper's reporting as 'dishonest and mendacious'. Columnists in broadsheet newspapers have themselves called for a degree of responsibility to be exercised by their colleagues - 'dog does eat dog' after all.

The Parliamentary Bureau has carried out its entirely novel, but vital, task of programming business with an increasingly sure touch. For example, the situation of the three individual Members, MM Canavan, Sheridan and Harper, was brought to the attention of the Bureau in regard to their membership of committees, and to keeping them informed as to its work. In both cases satisfactory compromises were reached: on 17 June the Executive's business manager, Mr Tom McCabe, announced that individual Members should have places on committees 'for all time'[1], having on 8 June announced that each of the three MSPs could sit on one committee[2]. On the same day it was confirmed that the Bureau briefed the three individual Members informally on its work, thus practising openness and an inclusive approach [3]. On 10 September the Presiding Officer, Sir David Steel, was reported as agreeing with members that technical help in drafting non-Executive Bills should be given by Parliament, a matter on which the CSG report is silent.

An encouraging example of co-operation between the parties to achieve an agreed objective was the conduct of proceedings on the Mental Health (Public Safety and Appeals) (Scotland) Bill. The Bill was rendered necessary by a decision in the courts that certain categories of patients in the State Hospitals suffering from a mental disorder might have to be released, even though they might constitute a danger to the public, by reason of a lacuna in existing legislation.

Using the emergency procedures set out in the Standing Orders, the Bill was considered at Stage 1 (consideration of principle) on one day; stages 2 (committee stage) and 3 (passing the Bill) occupied a further half day. The

Minister in charge of the Bill acknowledged the help that other parties had given, and the constructive way in which they had approached it[4].

The Principles in Practice: Leaning Towards Westminster

Speaking at a conference at the University of Dundee on 24 September 1999, John McAllion, MP and MSP for Dundee, East, sounded a clear note of alarm about the work of Parliament's committees. Reinforcing his speech by an article in the Herald newspaper on the following day, Mr McAllion contrasted the practices of Westminster committees with the ideals to be aimed at by their Scottish counterparts. But the latter were in danger of falling short of these ideals by reason of lack of funds to sit outside of Edinburgh; the workload on MSPs, and the workload of committees. He concludes that 'too much is being expected in too little time from committees'.

Mr McAllion also criticised the strict whipping foreseen for committee members during consideration of legislation proposed by the partnership parties in government; and the formation of party groups in committees. These latter two developments are probably unavoidable, at least in the early years of a Parliament : both were strongly reinforced, for example, in committees of the European Parliament after the first direct elections in 1979, having existed somewhat tentatively before then.

An apparent lack of funds for travel must be addressed in the long run, but in the short run it can be addressed by the convener, reporter and one representative from each party representing the committee. As regards the workload of MSPs, who sit normally on two committees, they must employ assistants to support their work and make use of information and communications technology to the maximum. It may be that committees are at present attempting - understandably - to undertake too many functions. Some prioritisation would seem desirable, otherwise 'the best becomes the enemy of the good'.

Another difficulty dogging committees at present is that three committee conveners act also as spokespersons for their party in Parliament and outside. Given the multi-functional character of the committees, which includes the need for conveners to be impartial in chairing inquiries into Executive policies and administration, it would seem more in keeping with key principles two and three if conveners did not speak for their party.

This point was forcibly argued by the First Minister, Mr Donald Dewar, in delivering a public lecture at Haddington on 9 November (Dewar). Mr Dewar went on to criticise 'a tendency to think that the key power of committees is to summon a minister to give evidence'. He suggested that committee agendas should not be dominated by headlines in the media arising from ministers' evidence to committees.

While this may be a moot point, Mr Dewar surprised some in his audience by trying to fly a kite which had hitherto seemed firmly grounded. He stated :

> Some will argue that reform of the House of Lords gives an opportunity to bind Scotland to the UK by giving the second Chamber the power to review Scottish legislation". It would allow a second opinion to be taken and give a strong justification for a Scottish input to the Lords.

Mr Dewar then however expressed the belief that 'an effective committee system is the best way of examining legislation and testing the Executive's actions'.

One of the leitmotivs of the work of the Scottish Constitutional Convention and the CSG was rigorously to exclude the House of Lords from any role in the Scottish legislative process. Yet only six months after the first meeting of Parliament, the First Minister appeared to be dragging into the democratic and modern Scottish legislative process the most unrepresentative body in the British constitutional structure.

The First Minister's kite met a razor-sharp response from the Presiding Officer of Parliament, Sir David Steel, who stated:

> Our [Scottish] legislative system is superior in its structure with its pre-debate examination, and I would regard interference by their Lordships to be contrary to the entire spirit of the Scotland Act, and unnecessary from what I have seen of the Parliament and its committees in action so far (Steel).

Sir David also rejected Mr Dewar's comments about hasty summoning of ministers before committees and defended the general approach of committees to their tasks at a very early stage in Parliament's life.

Yet another example of members being unable to resist the temptation to persist with Westminster methods and approaches broke surface in November. A UK minister, the deputy Secretary of State for Scotland and a Labour constituency MSP in the West of Scotland both attacked a Scottish Nationalist MSP, who holds a list seat in the same area. The Labour members claimed that the SNP list member 'had no mandate to negotiate anything' on behalf of the local electors, and that her views should be 'disregarded'. The Presiding Officer told the minister and the constituency MSP that -

> In the Scottish Parliament there is no distinction between those elected on a constituency, and those elected on a regional, basis. Both have equal mandates from the electorate [5].

This statement was followed shortly afterwards by a meeting between the Presiding Officer and one of his deputies with members representing the parties in Parliament to seek their agreement to a set of principles governing relations between constituency and list MSPs.

It may be noted that in the Bundestag, to which members are elected by a system similar to that used in Scotland in May 1999 no distinction is made between the two categories of member. This applies also to German members of the European Parliament.

The final evidence of an over-the-shoulder return to Westminster practices consists of an astonishing document, featured by the Herald newspaper, drafted by an official of the Executive Secretariat in the Scottish Executive. The document is entitled 'Dealing with the Scottish Parliament: Situation Report' and is dated 17 November 1999. A final draft of guidance on Executive evidence to Parliament Committees is referred to. The report continues:

> To be finalised this requires an agreement on certain aspects with the Parliament, and this will take a little time. In the interim, however, *colleagues should proceed as if the rules had already been promulgated* (Present author's emphasis).

Further paragraphs envisage restrictions on MSPs putting Parliamentary questions, and a 'Draft protocol for Clerks of Committees' dealings with the Scottish Executive' proposes six restrictions on the work of committees in relation to the Executive. These are based closely on rules forced upon Select Committees of the House of Commons by the UK government departments some years ago. This is the most flagrant attempt of the Scottish Executive to date to enforce Westminster rules on Holyrood.

Two final points remain to be made, arising from this document. The first is the unacceptable arrogance of officials' attitude to the Scottish Parliament, which flies in the face of the CSG's principles. The second is of more import: the principal challenge to the success of the Parliament is in the traditional attitude of the former Scottish Office, now the Scottish Executive. The Parliament will have to re-educate officials into the realities of 21st century Scotland, where the mantle of authority has passed to the people, represented in the Parliament, with which ministers and officials have to share power, and through which they are accountable to the electors.

The process of re-educating officials will take years rather than months. During it, Parliament may lose some battles. But unless it persists, it will lose the war, and the aspirations of the people, and of those devolutionist leaders who are genuine democrats, will be blighted. This single issue will weigh heavily in the judgement as to whether the Scottish Parliament will be a model for democracy or a 'mini-Westminster'.

Notes

1   Official Report, 17 June, col. 623.
2   Ibid, 8 June, c.263.
3   Ibid, 8 June, c.273.
4   Ibid, 8 September, c.271.
5   Herald, 12 November, 1999.

References

Dewar, Donald, First Minister. The John P. Mackintosh Memorial Lecture, Haddington, 9 November 1999.

Scottish Constitutional Convention (1995), *Scotland's Parliament, Scotland's Right*. The Scottish Constitutional Convention 1995.

Scottish Office (1997), *Scotland's Parliament* , The Scottish Office 1997

Scottish Office (1998a), *Shaping Scotland's Parliament* Report of the Consultative Steering Group, 1998 The Scottish Office, 1999, section two.

Scottish Office (1998b), Report, section one, para. 7.

Scottish Office (1998c), Report, section two, para 17.

Standing Orders (1999),   Statutory Instruments 1999 No. 1095. Constitutional Law. The Scotland Act 1998 (Transitory, Transitional Provisions) (Standing Orders and Parliamentary Publications) Order 1999.

Steel. Rt Hon. Sir David Steel, Presiding Officer. The Brough Lecture, Paisley University, 15 November 1999.

3   Some Constitutional Aspects of Devolution

JAMES G. KELLAS

Devolution is essentially a British invention. It first surfaced in the 1920s as 'federal devolution' and was applied to Northern Ireland between 1920 and 1972. No other country had devolution at this time, but several were federal. Federalism, however, came in many forms, notably in the U.S., Canada, Australia and Switzerland. By the time Scottish devolution was established, there were many more federal countries, such as Germany, Austria, India, Brazil, as well as some failed federations, such as the U.S.S.R., Yugoslavia and (briefly) the West Indies (1958-62), Rhodesia and Nyasaland (1953-63) and East Africa (Kenya, Tanganyika, Uganda) (1963-67); East African Community, 1967-77). There is also devolution in N. Ireland ('suspended' in 1972, but re-established in December 1999, then suspended again in February 2000 and re-established May 2000).

The causes of success and failure in federalism and devolution are not understood. Federalism was rejected for Britain by the Kilbrandon Commission and by successive governments (but not by the Liberal/Liberal Democrat Party). The reasons for that rejection are also obscure, but it was often said that the British state could not be a federal one mainly because England was too big and/or not ready for it.

Scottish devolution in its contemporary form was invented by the Kilbrandon Report in 1973 then in the Scotland Act in 1978. The Scottish Constitutional Convention reinvented it in the 1990s, and Labour moved very fast this time to legislate in 1997. No one knows yet whether this is a successful form of constitutional reform or a failure. That will depend on two factors: the wishes of the Scots, and the responses of the English. It will be noted that already 'British' has been airbrushed out in this account. That is because 'British' Government has ceased to be an issue: only Scottish government remains to trouble us.

Despite what was said by Labour politicians in 1997 at the time of the Scottish Referendum, devolution is not 'the settled will' of the Scottish people. In fact, the Scots are divided on whether devolution or independence is desired, sometimes (according to opinion polls) in the ratio 50:50, but more likely around 60:40 or 70:30 when the option of devolution is included along with independence. What is worrying for the British state, however, is the fluidity of this opinion, since poll responses have fluctuated not only according to the questions asked but from month to month.

The devolution settlement, intended for Scotland, Wales and Northern Ireland only, has in fact put the cat among the pigeons for England. England requires recognition in the British state, in a way that has never existed before. The focus of that recognition is the House of Commons rather than a separate English Parliament. That is because the English have always viewed the British Parliament as the English Parliament, and the British Constitution as the English Constitution.[1] (Bagehot). But that won't do when the Scots, Welsh, and Irish reject that construction. It is impossible to convert the British Parliament into a true English Parliament, however. The English will have to accept a parliament with Scots, Welsh and Northern Irish MPs voting on purely English matters. But there is nothing new in that, it is just that some do not want to accept it anymore.(The Scots came to that conclusion about English MPs voting on purely Scottish matters in the devolution Referendum in 1997). The Queen's Speech on 17 November 1999 contained the promise of 11 Bills out of 33 which did not apply at all to Scotland, since they covered devolved matters. In some of the other so-called 'U.K.' Bills there were matters relating to subjects which had been devolved to the Scottish Parliament, such as local government functions and agriculture.

Already, in the Chancellor of the Exchequer's pre-budget speech on 9 November 1999, it had been announced that the proceeds of the fuel and tobacco duties would be partly 'ring-fenced' to be spent on roads and the Health Service respectively. But these functions are devolved, and such spending would need to be authorised by the Scottish Parliament, even if the proceeds of the duties would come to Scotland automatically via the Scottish Consolidated Fund and the 'Barnett Formula', which allocates to Scotland a fixed percentage of the amount spent in England on the equivalent of the devolved services. The Scotland Act and the Scottish devolution settlement generally occupies an uneasy space between federalism and unitary government. While the British Parliament denies itself day-to-day power to pass Scottish Bills or question Scottish policies, it retains legal 'sovereignty' in the Scotland Act (Section 28 (7)), all the finance except for a possible variation in personal income tax of up to 3P in the £, and social security and the more important economic policies. A host of 'concordats' between Scottish and UK ministers has been made, no doubt with the intention to 'bind Scotland to the U.K.', in First Minister Donald Dewar's phrase[2]. While the language of the Concordats and of the accompanying *Memorandum of Understanding* between the U.K., Scottish and Welsh Governments is one of consultation and agreement, the outcome if no such agreement is reached is unclear. There is a scarcely veiled assumption that the U.K. could and would use its sovereignty (or more politely, its 'Lead' position in meetings) to get its way. There is even some legal opinion that these Concordats are justiciable, but who would benefit from that is not clear. An indication of how this binding of Scotland to the United Kingdom was given by Gordon Brown, Chancellor of the Exchequer, on 1 December 1999. Ministers of the Scottish Executive were to sit in 'joint-action committees' alongside members of the British Cabinet. According to a source close to the Chancellor quoted by the *Scotsman* on 29 November, 'it is time to move from emphasising the distinctiveness of separate institutions to showing how they can work together'. The policies involved included pensioner poverty, child poverty and development of a knowledge-based economy. It is clear that London has been worried that Edinburgh might get out of step in the programme of the Labour Government, but it can also be interpreted as an enhanced input of the devolved Executive into the U.K. policy making in reserved matters, of which social security and the economy loom large. The place of the Scottish Secretary (a U.K. Cabinet minister is cast into some doubt by these arrangements, however, as he was to play the part of link between Edinburgh and London in the British Cabinet. Now the Scottish Executive will have direct voice in British decision-making).

This can also be seen in the arrangements of the devolved civil service. The Scottish Executive's civil service is part of the UK Home Civil Service, which continues to regulate the entire administrative structure in Britain, including Scotland. No doubt, the 'ethos' of one service, and the frequent interdepartmental meetings will also 'bind Scotland to the U.K.'

Even the Clerk of the Scottish Parliament has come from the civil service (the old Scottish Office). This seems to offend against the British Parliament's practice of not recruiting its staff from the executive. But the only alternative was to recruit from the British Parliament (some Clerks and other staff did come from the House of Commons).

First Minister Donald Dewar was until May 1999 a U.K. Cabinet Minister, and John Reid, the Scottish Secretary, occupies a baffling position as U.K. Cabinet minister with hardly any executive functions. In the Scotland Office Website Homepage [3], it is stated that the department will 'promote devolution' and act as an 'honest broker' in disputes between the Scottish and U.K. Governments. It is strange that an 'honest broker' is actually a member of one of the parties' teams. In practice, Reid and his one other minister, Brian Wilson, spend most of Scottish Question-Time in the House of Commons defending their very presence in the U.K.

Government against attacks by sceptical Conservative and Nationalists who believe that the Department is unnecessary after devolution.

Lawyers are everywhere in the new Scottish political system, and a new 'Dundas Despotism'[4] is on the way, albeit mostly for Scotland's own political system, not often as a feed-in to the House of Commons. That is because the Scotland Act puts the Scottish Parliament and Executive under judicial control, unlike the 'sovereign' U.K. Parliament and Government. Not only can judges rule Acts of the Scottish Parliament and decisions of the Scottish Executive *ultra vires*, but even the procedures of the Parliament have come before the Court of Session on two occasions.[5] A U.K. Court, the Judicial Committee of the Privy Council, is the final court in 'devolution cases', and in time no doubt a legal canon will develop comparable to that of the Supreme Court of the United States in federal-state relationships but with the important difference that no Court can overrule the British Parliament and Government. A new British law officer, the Advocate General for Scotland and a team of around 30 lawyers are employed in the Scotland Office, partly to assist in drafting U.K. Bills which have Scottish sections, but also to act as prosecutor of the Scottish Parliament with regard to *ultra vires*. Even the English Attorney General has the right to haul the Scottish Parliament before the Judicial Committee of the Privy Council. (Scotland Act, Sections 32 to 34).

All this looks unstable, and a recipe for independence as it becomes clearer that 'binding Scotland to the U.K.' in this way will negate the purpose of devolution, as perceived by the Scottish people.[6]

Notes

[1]   W. Bagehot, *The English Constitution* (1867).

[2]   Donald Dewar, in his John P. Mackintosh lecture of 9 November 1999, here referring to the use of the House of Lords as a revising chamber for Scottish legislation. *Scotsman*, 10 November 1999.

[3]   www.scottishsecretary.gov.uk

[4]   Henry Dundas, 1st Viscount Melville (1742-1811), was Lord Advocate and 'Scotland's Manager' for the U.K. Government. He was a lawyer in a position of supreme political power in Scotland in the era of the unreformed House of Commons. He dispensed favours to electors in Scotland in return for their votes, and 'managed' the affairs of Scotland in London and Edinburgh. The 'Dundas Despotism' is the phrase applied to his rule, and to those of lawyers in Scotland at the time.

[5]   The first occasion was a case brought by the *Scotsman* on 4 October 1999 concerning the Parliament's Standards Committee decision to hold a meeting in private. The *Scotsman* considered that this was contrary to the Parliament's Standing Orders. The Court of Session would have taken this case up, but the *Scotsman* dropped it on 17 November, as by that time the Committee had gone

public. The second case concerned an alleged conflict of interest of an MSP, Mike Watson, with regard to his introduction of a Bill banning foxhunting, since he had been supported by anti-hunting Organisation. On 26 November 1999, Lord Johnston dismissed the case on the grounds that the matter was for the Parliament to decide. Although these cases have not decided against the Parliament, the principle apparently remains that the Parliament's procedures are justiciable in some circumstances.

6

For these perceptions and expectations of devolution in surveys, see A. Brown, D. McCrone, L.Paterson and P. Surridge, *The Scottish Electorate. The 1997 General Election and Beyond,* especially ch.7 (Basingstoke: Macmillan, 1999).

4 Scottish Politics After the Election: Towards a Scottish Political System?

ALICE BROWN

Introduction

> Scotland's already well established political distinctiveness will continue to develop as the first Scottish Parliament is elected and begins its proceedings (Dunleavy et al, 1997).

This quotation sums up succinctly the focus of this chapter, namely the extent to which the distinctive nature of politics in Scotland has developed following the first Scottish parliamentary elections held in May 1999. Further it relates very directly to whether we can observe a distinct Scottish political system. At the time of writing, less than one year after the elections, it is difficult to give more than a preliminary assessment of the impact of the establishment of the Scottish Parliament. But one approach is to look back to the campaign for constitutional change in order to assess its influence on the system that is evolving and to begin by asking how we would have characterised Scottish politics in the past. It is argued here that political developments during the 1980s and 1990s helped shape the type of parliament that has been set up in Scotland and fuelled expectations of a different type of politics that could be distinguished from the Westminster model. The setting up of a new institution was perceived as a unique opportunity to build a more democratic political system and more inclusive political culture. Whether these aspirations have been achieved is, of course, still open to question.

Scottish Politics Pre-1990

With notable exceptions, very few published books in political science addressed the specific topic of Scottish politics in the past. Publications on British politics at best included a chapter on Scotland but more often discussed and analysed the British political system as if this was an unproblematic and homogenous concept. But, even before the setting up of the Scottish Parliament, distinctive elements of politics in Scotland could be identified. For example, with its own nationalist party, the Scottish National Party, Scotland had a four-party system, and with their own histories and ways of operating, the political parties were not just smaller versions of their British counterparts. In this four-party system and operating under the first-past-the-post electoral system, it was the Labour Party that was the main beneficiary and enjoyed hegemony at all governmental levels - local, Westminster and European (Brown, et al, 1996 and 1998).

Commentators also noted the growing divergence in voting patterns between Scotland and England. From a position of gaining over 50% of the vote in Scotland in 1955, the Conservatives witnessed a steady decline in their support to an all time low in 1987 when only 10 Conservative MPs were elected to represent Scottish constituencies, causing difficulty in filling all the ministerial places in the Scottish Office. Rather than the main party competition being between the Conservatives and Labour, as in England, increasingly the key rivalry was between the Labour Party and the SNP (Brown, et al 1999).

In addition to these party and electoral factors, some policy divergence north and south of the border was also evident both in terms of the policies themselves and the process by which policy was made. As Paterson (1994) reminded us, even without a parliament, Scotland enjoyed a level of autonomy over its own affairs, especially as it maintained its own legal, education, church and local government systems following the Treaty of Union in 1707. Such policy autonomy and separate systems were enhanced by other distinctive Scottish institutions including the media, business organisations, trade unions and voluntary sector. Again these were not just 'branch plants' of London based bodies but operated within a Scottish political context which allowed a role for Scottish civic society and the development of a citizenship based on notions of demos as distinct from ethnos (Meehan, 1993). Similarly the growth of nationalism in political parties - not just within the SNP - and in other institutions in Scotland was described as civic nationalism rather than ethnic nationalism (McCrone, 1992).

Crucially it was also the role of the constitutional question which provided a distinctive element of politics in Scotland, a role which we will discover increased over the 1980s and 1990s. In sum, all of these elements fostered an economic, social and political system which could be distinguished from politics in England. Writing in 1973, James Kellas talked about 'The Scottish Political System' in his celebrated book of the same title. As Kellas (1989) acknowledged, this was not a description that was accepted by everyone and he was challenged on the grounds that Scotland lacked the key ingredients required to qualify as a political system of its own. Kellas cited John Mackintosh's view that Scotland could not be said to have its own political system on the grounds that it did not have a focal point where various Scottish pressures met; Richard Rose who argued that Scotland could more accurately be described as a 'sub-system' because it was dominated by a British government which itself was dominated by English politics; and others including Michael Keating and Arthur Midwinter who preferred to talk in terms of policy networks operating in Scotland and stressed that ultimate authority rested in London.

In reply, Kellas (1989) argued that to a large extent these critiques suffered from a similar flaw in that they were based on a top-down view of Scottish politics. The question was posed as to whether the distinctiveness of Scottish politics could be most clearly illustrated as a bottom-up movement. Kellas concluded that the evidence from his book pointed to 'the existence of a Scottish political system comprising a large number of actors engaged in a wide range of political actions' (1989: 256).

It is fair to conclude that although the extent to which politics in Scotland could be defined as a Scottish political system was open to challenge in the past, most commentators did acknowledge at least some elements of difference or distinctiveness.

Changes Post-1990

There are a number of factors in the 1990s which have left a legacy and have influenced the nature of post-devolution politics in Scotland. The first is the broad-based campaign for constitutional reform which gathered steam over the period. Significantly this did not just include the political parties in Scotland. It was at times fronted by other institutions in Scottish civic society, lending weight to Kellas' notion of a bottom-up movement. The involvement of different groups and organisations has had implications for perceptions of 'ownership' of the new parliament and for aspirations surrounding the way in which it conducts its affairs.

Secondly, the campaign for constitutional reform in Scotland was in part a reaction to what has been described by some as a 'democratic deficit'. In the process of arguing for a Scottish Parliament, strategic alliances were forged across the political party divide and between parties and civic organisations. Such alliances helped to influence the debate on the introduction of proportional representation and a vision of a different type of politics in Scotland. This vision of difference was based on a critique of the Westminster style of government which was seen as remote, highly adversarial, centralised and secretive, unrepresentative of the general population, and which operated in ways that presented unnecessary barriers for the participation of people with family responsibilities or from remote parts of Britain. The Scottish Constitutional Convention and the Scottish Civic Assembly were just two of the vehicles where plans were made for a Scottish Parliament which operated on different lines from the Westminster model (Brown et al 1996 and 1998).[1] Therefore, when the Convention published its final report, *Scotland's Parliament, Scotland's Right*, on St Andrew's Day 1995, it represented a broad-based and collaborative approach to setting out a scheme which included proposals for a parliament with tax-varying powers and elected by a more proportional electoral system with gender balance. The scheme was endorsed by the two main political parties in the Convention, the Scottish Labour Party and the Scottish Liberal Democrats, as well as other small parties, the trade union movement, the voluntary sector, the churches, local government, women's organisations and other groups in Scottish civic society. Reflecting the way in which agreement on a scheme was reached, the Convention's report set out its aspirations for the future in the following terms:

> From this process we have emerged with the powerful hope that the coming of a Scottish Parliament will usher in a way of politics that is radically different from the rituals of Westminster; more participative, more creative, less confrontational - a culture of openness which will enable the people to see how decisions are being taken in their name, and why. The Parliament we propose is much more than a mere institutional adjustment. It is a means, not an end (Scottish Constitutional Convention, 1995).

The next significant step on the road to the Scottish Parliament was the election of the Labour government in May 1997 with an overwhelming majority (see Table 4.1). The election results in Scotland demonstrated clearly how the FPTP electoral system operated to the benefit of the

Labour Party in particular (Brown, 1997). With less than 46% of the vote, Labour received around 78% of the seats. As in Wales, the main casualty of the elections was the Conservative Party who, against most predictions, lost all their Westminster seats. With the effective removal of the party who had defended the status quo and warned against the establishment of a Scottish Parliament, the way was clear for the implementation of plans for constitutional reform.

Table 4.1 1997 General Election in Scotland

Party	% of Vote	No. of Seats	% of Seats
Labour	45.6	56	78
Lib.Dems.	13.0	10	14
SNP	22.1	6	8
Conservative	17.5	0	0
Others	1.9	0	0
Total:		72	

Dispelling doubts that it may not keep its pre-election promises, the Labour Party moved quickly to publish its White Paper on Devolution in July and to hold the two-question referendum in September 1997. Crucially, the SNP combined with Labour and the Liberal Democrats and others players in the home rule campaign to help deliver a double yes vote.[2] With the endorsement of majority support for a parliament with tax-varying powers, the scene was set for the publication of the necessary legislation. The work carried through the Scottish Constitutional Convention and their key recommendations were reflected in the Scotland Act, which in spite of some controversy over aspects of powers retained at Westminster, contained no real surprises and was widely endorsed. The spirit of alliance building was continued with the establishment of the Consultative Steering Group in 1998 involving representatives from the four main political parties and others from civic society. The group was given the remit of recommending standing orders and procedures to the new politicians once elected to the parliament in 1999.

The key point being made in the above discussion is the significance of these developments in terms of potential impact on the scheme for the Scottish Parliament and politics after the first elections. Given the support for the parliament in the referendum, the legitimacy of the project was secured with even opponents of the parliament in the Conservative Party and the business community accepting the result and pledging to work constructively to make the parliament work. The specific scheme proposed and the work of the Consultative Steering Group also continued the inclusive process of decision-making on constitutional matters which was not confined to the party in government, or indeed to the political parties as a group, but encompassed other key stakeholders. To a large extent, therefore, by the time the elections for the new parliament were held in May 1999, much of the constitutional debate had already taken place and a broad consensus over parliamentary arrangements had been reached. Understandably, the alliances that had been forged did not extend to other aspects of party politics nor did it prevent fierce competition for seats in the new parliament.

The First Scottish Parliamentary Elections - May 1999

The May 1999 elections have been described and analysed by other authors in this volume (see Chapters 5 and 6) and elsewhere. As Peter Jones (1999) has stated the word 'historic' was an overused term in the elections but was unavoidable given the circumstances. Scotland was voting for a new institution, with a new electoral system, and for new MSPs elected under new selection processes and mechanisms for gender balance. It was also anticipated that the newly elected politicians would operate under the new procedures recommended by the Consultative Steering Group including a new committee structure and policy process. The questions asked were whether such new arrangements would result in a new political culture in Scotland and a different way of conducting politics, or would it be a new institution with rather old politics.

As anticipated prior to the election, no single party won an overall majority and people in Scotland used their two votes strategically and in ways which were not entirely in line with voting patterns for Westminster elections. Although the introduction of the new AMS (see p. 52) for the elections did not achieve exact proportionality, as Table 4.2 illustrates there was a closer relationship between the percentage of votes cast and percentage of seats gained by the political parties, and the Conservative Party gained seats in a parliament they had opposed and under an electoral system they had rejected. As Lynn Bennie discusses (Chapter 6), the new electoral system also provided the opportunity for representation from smaller parties. Campaigns to improve the representation of women also had an impact. For example, the Labour Party introduced the policy of 'twinning' constituency seats with the result that they achieved 50:50 representation of men and women. As Table 4.3 shows, the SNP were also successful in returning a large proportion of women MSPs by placing them high on the regional party lists (Brown, 1999).

Table 4.2 Scottish Parliamentary Elections - May 1999

Political Party	Constituency Vote (%)	Regional Vote (%)	No of Seats Const./ Reg.		Total Seats (%)
Conservative	15.6	15.4	0	18	18 (14)
Labour	38.8	33.8	53	3	56 (43)
Lib.Dems	14.2	12.5	12	5	17 (13)
SNP	28.7	27.0	7	28	35 (27)
Others	2.7	11.4	1	2	3 (2)
Totals			73	56	
Overall Total:			129 seats		

Table 4.3 Gender Composition of the Scottish Parliament - May 1999

Political Party	Elected MSPs Women Numbers	Men	Elected MSPs Women Percentage	Men
Conservative	3	15	17	83
Labour	28	28	50	50
Lib.Dems	2	15	12	88
SNP	15	20	43	57
Others	0	3	0	100
Totals	48	81	37%	63%

With the elections over, the first meeting of the parliament was on 12 May 1999 and the new politicians began the process of electing their Presiding Officer and Deputes before moving on to elect the First Minister, Donald Dewar. With no overall majority, speculation surrounded whether the Labour Party would attempt to go it alone or would, as predicted and discussed prior to the election, form a coalition government with their partners in the Scottish Constitutional Convention, the Scottish Liberal Democrats. After long negotiations, a partnership government was agreed on 14 May and the first Minister went on to appoint his Cabinet of 11 Ministers (9 Labour and 2 Liberal Democrats) assisted by 11 Junior Ministers (also 9 Labour and 2 Liberal Democrats).[3] Departments within the old Scottish Office, now re-named the Scottish Executive, were restructured to reflect the new portfolios including Justice; Children and

Education; Enterprise and Lifelong Learning; Finance; Health; Rural Affairs; Social Inclusion, Local Government and Housing; and Transport and Environment. In the parliament itself eight subject committees were set up to reflect the new departments in addition to eight mandatory committees, namely European, Equal Opportunities, Finance, Audit, Procedures, Standards, Public Petitions and Subordinate Legislation.

As this short description illustrates, from the very beginning procedural differences from Westminster were evident as the new parliament began to run its affairs on the lines proposed by the Consultative Steering Group. A significant feature of the difference was the role anticipated for parliamentary committees. Unlike the Westminster model, parliamentary committees have both standing and select functions and as well as scrutinising and monitoring government legislation, they can hold their own enquiries and take evidence and advice from different quarters. As recommended by the Consultative Steering Group they can also initiate legislation. Another important feature is that membership of the committees reflects the party balance and the effects of the more proportional electoral system, and not all conveners are from the coalition parties. This marks a radical departure from the type of system operating in the House of Commons. It was also recommended that the committees should have cross-cutting functions and should meet in different parts of Scotland to consult and engage as many people in their deliberations, and there is already evidence that some committees are experimenting in these ways. However, more radical proposals by the Consultative Steering Group that some committees should be based outside Edinburgh and should be able to have non-MSPs have not yet been implemented.

In looking at what the parliament actually does, another point of comparison with Westminster can be drawn. The Scottish Executive embarked on an ambitious legislative programme including eight bills proposed for the first year covering aspects of education, transport, land reform, the feudal system, national parks, local government, incapable adults and finance and auditing. Provisions also exist for emergency legislation and for the introduction of Private Member's Bills. This contrasts dramatically with the time allocated in the past to Scottish legislation in the House of Commons.

Less positively, the early weeks and months of the parliament were dominated by adverse media attention surrounding disputes over the allowances paid to MSPs and what is known as 'Short' money, that is the money paid to the opposition parties to allow them to scrutinise the government. Controversy continued with the so-called Scottish lobbygate affair in the autumn of 1999 and the first big test for the coalition government over tuition fees for students in higher education. Negative media attention continued over the departure of Special Advisers and with the dispute over the repeal of Clause 28. An ongoing source of division relates to the site and the cost of the new parliament building at Holyrood. All of these more negative aspects have detracted from a more balanced and thorough examination of the working of the parliament and its new parliamentary and policy making processes, and an assessment of the Scottish Executive and analysis of Scottish politics more generally.

Scottish Politics after the Election

Clearly there are inherent tensions in the new political settlement in Scotland as in any political arrangements where power is devolved. The first relates to the funding of the parliament. Although there is the power to vary taxation by 3p in the £, the Labour Party made it clear prior to the election that they would not exercise this power if they formed a government in 1999. The main source of funding remains the same as that which operated under the Scottish Office based on a UK allocation under the Barnett formula. Such an arrangement could lead to problems not only with regard to the autonomy of the Scottish Parliament to raise a significant amount of its own funding but in relation to other parts of the UK, especially regions of England, where the amount allocated to Scotland has been, and is likely to continue to be, contested.[4] Yet, at the time of writing, the financial constraints and the squeeze on Scotland's share of public spending, have not as yet resulted in the type of political problems for Labour in Scotland that were anticipated prior to the election.

Related to tensions over the distribution of funding, is the division of powers between the different tiers of government. British politicians in general have yet to come to terms with living in a multi-layered democracy where many aspects of policy are determined at the European level and where local government is keen to regain powers they lost under the Conservative governments in the 1980s and 1990s. In Scotland, the new parliament has its own European Committee, even although Europe is a reserved power; and the Macintosh Commission which examined the role of local government after the establishment of a Scottish parliament argued strongly on the grounds of parity of esteem for different governmental levels. As John Fairley discusses in Chapter 7, the relationship between the parliament and local government has still to be worked out. The publication of Concordats covering the different tiers of government provide a framework for the different relationships involved but much will evolve in the ongoing process of reform.

Another potential tension and difficulty for the future is the extent to which the parliament will adhere to the key principles set out by the Consultative Steering Group, namely those of power-sharing, accountability, access and participation, and equal opportunities. Precisely what the implications of power-sharing and greater access and participation are for the relationships between the executive and the parliament, the parliament and the civil service, and the parliament and civic society are as yet unclear. While the aspiration to develop a different political culture and system are real and strongly felt by some of the new politicians, the constraints are also very considerable. This also raises issues about the relationship between the different type of MSPs. Some of the new politicians are very new indeed to political life in terms of parliamentary office but some are currently holding dual mandates with their Westminster seats. The electoral system itself brought politicians into the parliament elected under different parts of the process and the relationship between constituency MSPs and list MSPs has yet to settle down. Much depends too on the party rivalries in particular parts of Scotland. Members elected under the list system may be especially anxious to please their party leaders and the extent to which the parliamentary committees provide a real challenge to the executive may be restricted by conveners keen to 'please the Minister' with a view to future career prospects.

And last, but not least, there is the whole question of coalition politics within a multi-party system. On the one hand there is a Labour Party in Scotland that is not accustomed to sharing power. Although they do not hold an overall majority in the new parliament, they still have considerable proportion of the seats. On the other, their coalition partners, the Liberal Democrats, have been accused on occasion of forgetting that they are now a party of government and not the opposition. As indicated above, the first big test for the coalition came in the form of the difference of opinion over the abolition of tuition fees for students in higher education. The Scottish Liberal Democrats continued to espouse their pre-election promise to abolish fees and they received support for this policy from all the other parties in the parliament with the exception of their coalition partner. With the official position of the Scottish Labour Party being to maintain fees, the potential for a difficult political situation was set, so much so that some commentators predicted the collapse of the coalition by December 1999. Such a crisis was avoided by the setting up of a Commission of Enquiry under the chairmanship of Andrew Cubie. The Committee operated on lines very much in tune with the work of the Consultative Steering Group and canvassed broad opinion throughout

Scotland. The Cubie Committee reported early in 2000 with recommendations that received widespread endorsement in effect ending tuition fees. Following a debate in the new parliament, the key findings of the report were accepted with modifications. This provided an example, however, that even as a smaller partner in the coalition, the Liberal Democrats could exert power in the context of multi-party politics. It was an important landmark also as it brought to the attention of politicians and others in England the fact that the Scottish Parliament could indeed make policy in areas that were distinct from those south of the border, even although some of Cubie's recommendations are restricted by the fact that social security remains a reserved power.

Towards a Scottish Political System?

Having charted some of the developments of the political process in Scotland both prior to and following the first elections to the Scottish Parliament we can now return to our original question - whether politics in Scotland are becoming more distinctive and can be described as a Scottish political system.

Party System

If we return to the categories we identified to describe Scottish politics prior to the election, we can observe that rather than the two-party system operating in England, Scotland can claim to have a multi-party system which includes a role for the smaller parties (see Chapter 6). The main electoral competition for Scottish parliamentary seats is between the Scottish Labour Party and the SNP, and in future elections the exact composition of the Scottish Parliament - and of course the Scottish Executive - may alter as new alliances are forged. While the AMS was designed in such a way that it was unlikely that the Labour Party would achieve an overall majority, this means, of course, that is also unlikely that any other of the main parties will be able to do so either at least in the foreseeable future. It also means that Labour may have to face the uncomfortable position of not being part of the government of Scotland in the future.

The new political context in Scotland has also increased pressure on the political parties to emphasise the distinctiveness of their

Scottishness. For Westminster elections, Labour can claim to be the party most able to speak for Scotland in London. But in Scotland the big question is who speaks for Scotland in Scotland. All of the parties have undergone some re-examination of their role and strategy following the British general election in 1997 in order to prepare for the first Scottish elections in 1999. There are challenges for all the larger parties as Labour seeks to assert its image as Scottish Labour while at the same time maintaining compatibility with aspects of the programme for the Westminster parliament. The Scottish Conservative Party was quick to address its perceived anti-Scottish image and one of the ironies of politics is that it has sought to re-build the party through its representation in the Scottish Parliament. The success of the party in the Ayr by-election early in 2000 when it regained what was regarded as a Conservative stronghold seat provided encouragement for party managers that their strategy was paying dividends. The Scottish Liberal Democrats have well established Scottish credentials. Nevertheless, their new role as part of the governing coalition has presented new challenges for the party. Although not constrained in the same way as the others, the SNP re-examined its strategy and image following the Scottish parliamentary elections, even adopting some of the language and rhetoric associated with the Blair modernisation programme. In some ways it has a difficult task working positively in a devolved parliament while at the same time arguing for independence. But it does have the luxury if things go well to claim that it has played a key part in ensuring success. If things go badly and the parliament fails to meet expectations, it can continue to argue that independence in Europe is the only solution to Scotland's problems. The SNP's announcement that it will hold a referendum should it form a government after the next Scottish elections illustrates the shift in the party's strategy.

Electoral Divergence

Although with the large swing to Labour across Britain in 1997 the electoral divergence between Scotland and England was narrowed, there is evidence of differential voting patterns for the Westminster and Edinburgh elections as voters in Scotland, like voters in Catalonia, distinguish between elections for parliaments at different levels of governance. The results of the next British general election, which is expected to take place some time in 2001, will provide interesting evidence on which to judge emerging patterns of voting in a system which allows electors the chance to cast their votes in different ways under separate electoral systems operating in different parts of the country.

Public Policy

The level of policy divergence or distinctiveness in Scotland following the establishment of the Parliament could vary according to who holds office in Westminster and in Edinburgh. It may not always be the case that the same party is in power in both parliaments. As has been discussed above, however, there is already evidence that even with the same party in power, the Scottish Parliament is going its own way on policy matters. Apart from the example of tuition fees, the executive and parliament are making policy in other areas which meet the needs of conditions in Scotland and which they are keen to portray as Scottish solutions to Scottish problems. As described, the Scottish Executive has a comprehensive programme for legislation. The parliamentary committees are also finding their feet and exercising influence. The Justice Committee and the Enterprise and Lifelong Learning Committee are just two examples that have been cited to illustrate the impact of the new parliamentary arrangements.

Civil Society

As far as the role of civil society is concerned, they are working to play their role in the policy process in Scotland, both in respect of interaction with the Scottish Executive and with the parliamentary committees. The potential to enhance the scope of citizenship and to include new voices in this process is increased with the setting up of a Civic Forum in 2000. But as Grant Jordan and Linda Stevenson (Chapter 12) remind us the involvement of some interests raises other issues and questions for democracy and democratic participation. It is too early yet to judge whether recent developments will represent a continuation of what some regard as an inherently conservative and elite consensus, and provide greater space for a social conservatism that may have not found a vehicle for its expression in the past; or whether the existence of the parliament will offer the opportunity to set a more radical agenda where new and different alliances are forged between parliamentary backbenchers and others in Scottish civil society intent on pursuing a more progressive agenda. Therefore, whether civil society continues to play an essential oppositional role as it did prior to the establishment of the parliament is also open to question, and what that will mean in the new political context is also unclear.

Constitutional Question

The other distinctive feature that we identified to describe Scottish politics in the past was the role of the constitutional question. Will the new parliament be the 'settled will' of the Scottish people as John Smith described it or is it the beginning of a process which will continue ultimately to full independence? Again it is not possible to provide an accurate response to such speculation. It is clear that constitutional change is more than just an event with the setting up of the parliament but a process that will continue to evolve and could be seen as just another stage in the 'negotiated compromise' in the ongoing relationship between Scotland and England to which Paterson (1994) refers. In an ever changing Europe in which the whole concept of independence of individual states has a new meaning, it is also impossible to predict precisely where constitutional reform will end. It is not inevitable that it will mean full independence for Scotland, on the other hand as evidence from elections studies illustrate, this is a outcome which is not feared by most people living in Scotland (Brown et al, 1999). Further, whatever happens in Scotland will also be influenced by constitutional changes in others parts of the UK and whether or not the Good Friday Agreement is able to operate and the British-Irish Council becomes fully effective. In this whole process of constitutional change there are still far more questions than there are certain answers.

Gender Politics

We can also add another dimension to the picture that was not included in the past and that is the role of gender politics. Campaigns by women in political parties, trade unions, local government and the women's movement helped ensure that the representation of women became part of the package of constitutional change. The pressure to give women a greater say in politics and public life is not confined to Scotland and is part of a wider international trend. However, the debate took a particular form in Scotland because of the specific opportunities offered in setting up a new institution, under a new electoral system with new parliamentary arrangements. The political parties themselves also altered their selection processes in preparation for the first elections to the Scottish Parliament. Women activists seized the potential offered by such changes to make the case for a parliament with more equal representation and which operated in ways that did not present unnecessary barriers to the participation of people with family responsibilities. The forty-eight women were elected to the new parliament - some 37% - broke all records for women's representation in Scotland. While their presence initially attracted mainly negative reactions from the media, there is growing evidence that their contribution is being acknowledged and, in some cases, valued.

But the object of getting more women into the parliament was not just an end in itself. It was viewed as a means to other ends, namely to help improve the way in which politics is conducted and to have an impact on political decisions and on the priorities and focus of public policy. As recommended by the Consultative Steering Group the parliament has adopted so-called 'family friendly' hours of operation; and it has established an Equal Opportunities Committees. In addition an Equality Unit has been set up in the Scottish Executive and the policy of mainstreaming equality endorsed. A Women in Scotland Consultative Forum is another way in which the voices of women living in Scotland are being inserted into the consultative and policy processes. This has led some women's organisations to observe that they now have unprecedented access to ministers and policy-makers. More evidence will be needed, however, on which to judge the precise impact of these changes and the extent to which the participation of more women has 'made a difference' both to the political culture but also to policy outcomes.

Conclusion

We have discussed the initial impact of the Scottish elections on Scottish politics. The new parliament with its new electoral system and parliamentary arrangements have already resulted in a growing political distinctiveness in Scotland. An enhanced role for women in politics and for Scottish civil society are also evident, as are opportunities for different forms of participation and alliance building in the new political context. Elements of what some have described as 'new politics' can be identified although aspects of continuity and traditional political behaviour are also very much alive. Because of internal and external political forces, the longer term effects and impact of the Scottish Parliament are more difficult to judge. Nevertheless, we can return to Kellas' assertion about the existence of a Scottish political system with more certainty. It would be much more difficult for critics to argue that Scotland does not have its own political system. It certainly now has the key focal point that John Mackintosh thought was so crucial. It may be more difficult to convince Richard Rose as he may still want to argue that Scottish politics is still dominated by a British government which is in turn dominated by English politics, or indeed Michael Keating and Arthur Midwinter on the grounds that power still rests with Westminster. While it is undoubtedly the case constitutionally that sovereignty lies with Westminster, politically the balance of power has shifted and will continue to shift. When most people and commentators in Scotland talk about 'the parliament', they are mainly referring now to the Scottish Parliament. Thus 'the' has become shorthand for 'Scottish'. It is the House' of Commons that they now distinguish by prefixing it with its full title of the Westminster Parliament.

There is evidence to support the argument made by James Kellas (1989) that: 'There is no doubt that devolution or federalism would strengthen the Scottish political system and perhaps make it a different kind of political system.' There is also evidence to support the view that Scotland is moving away from the adversarial, zero-sum style of politics associated with Westminster and developing a more inclusive and plural political system and political culture. We can argue that after the first elections to the Scottish Parliament, Scotland does have its own clearly defined and evolving Scottish political system and that Scottish politics will never be quite the same again. No longer will it be tenable for books on British politics to be published with assume the homogeneity of the British state and the British political system. The establishment of the Scottish parliament together with the Welsh Assembly, the Northern Ireland Assembly, Regional Development Agencies in England and the London Authority with an elected mayor, together with constitutional changes to the House of Lords and other proposals, calls into question the way in which we understand, analyse and theorise the politics of these islands.

Notes

1 In the SNP separate constitutional plans were developed for an independent parliament of 200 members within the European Union.

2 Constitutional campaigners combined under the organisation Scotland Forward. Those who opposed constitutional change, mainly supporters of the Conservative Party and some business leaders, worked together in the Think Twice or No/No campaign. The result of the referendum was that more than 74% of people in Scotland who voted backed the idea of a parliament with almost 64% agreeing that it should have the power to vary taxation.

3 This represented a gender balance of 17 men and 5 women (77%:23%).

4 The debate concerning whether Scotland is a net beneficiary or net contributor to the UK Treasury is a long and ongoing one. It featured also in the elections for the London mayor in spring 2000 when one of the candidates, Ken Livingstone, claimed he would obtain a better allocation

for London and would attempt to reduce the amount of NHS funding going to Scotland.

References

Brown, A. (1997), 'Scotland – Paving the Way for Devolution?', *Parliamentary Affairs*, vol.50, no.4.

Brown, A. (1999), 'Taking Their Place in the new House: Women and the Scottish Parliament', *Scottish Affairs*, no.28.

Brown, A., McCrone, D. and Paterson, L. (1998, 1996), *Politics and Society in Scotland*, Macmillan, Basingstoke.

Brown, A., McCrone, D., Paterson, L. and Surridge, P. (1999), *The Scottish Electorate*, Macmillan, Basingstoke.

Dunleavy, P., Margetts, H. and Weir, S. (1997), *Devolution Votes: PR Elections in Scotland and Wales*, Democratic Audit Paper, no.12.

Jones, P. (1999), 'The 1999 Scottish Parliamentary Elections: From Anti-Tory to Anti-Nationalist Politics', *Scottish Affairs*, no. 28.

Kellas, J.G. (1989), *The Scottish Political System*, 4th edition, Cambridge University Press, Cambridge.

Meehan, E. (1993), 'Citizenship and the European Community', *Political Quarterly*, 64(2).

McCrone, D. (1992), *Understanding Scotland: the Sociology of a Stateless Nation*, Routledge, London.

Paterson, L. (1994), *The Autonomy of Modern Scotland* , Edinburgh University Press. Edinburgh.

PART II
THE 1999 ELECTION
AND POLITICAL PARTIES

5 How Scotland Voted in 1999

IAN CLARK

Introduction

This chapter considers the results from the inaugural Holyrood elections of 6 May 1999. In the space available it is, of course, impossible to cover everything in detail, so most of the analysis deals with aggregate results within the regional list areas or across Scotland as a whole, rather than individual results from the constituency battlegrounds.

A brief caveat on comparisons with the 1997 General Election is necessary. Although 71 of the 73 Scottish constituency seats in 1999 were fought on identical boundaries to those in effect in 1997, there are a number of factors that hamper direct comparisons. Firstly, while the first past the post voting system was retained for constituency seats in 1999, the ability of electors to cast a second vote in list seats may have encouraged tactical voting. Secondly, there was a drop in turnout from 71.8% in 1997 to 58% in 1999, and though this was better than recent European and local elections it was more than 10% down on the norm for a General Election. A change of that size clearly had an impact on the results, though one that cannot easily be measured. Thirdly, only 15 of the 72 Scottish MPs at Westminster stood for election to the Scottish Parliament, so there was little opportunity for incumbent parties to mobilise MPs' personal votes. Indeed, the relative anonymity of many of the candidates fielded by the four main parties may have contributed to the fall in turnout, and the comparatively better showing of minor parties and individual candidates in the regional lists. Finally, it is important not to forget that this is a completely new Parliament with a wide range of legislative powers, and that the political imperatives that led electors to choose a Westminster MP may be significantly different from those which influenced their votes for Holyrood. It is likely that after the next Westminster election, the number of Scottish seats will be reduced to match the ratio of seats to electors extant in England, as provided for in section 86 of the Scotland Act 1998. If so, there will be a commensurate reduction in Holyrood constituencies for the first Scottish general election after the Westminster seats change, and this would further complicate direct comparisons at constituency level between results from the two legislatures for at least a couple of elections.

The New Voting System

The additional member list system was used for the Holyrood elections. There were 73 single-member constituency elections (the former Orkney and Shetland constituency was split to give each set of islands its own seat), conducted by simple plurality, so that the candidate with most votes won, no matter how small the majority or share of the vote. These constituencies were grouped in eight regional areas, in which 56 'additional members' (seven from each area) were elected from party and individual lists. The regional party lists were 'closed'; in other words, prepared by the parties, who decided the order in which names appeared on the list (and thus the order in which they would be elected), thus voters were unable to influence that order. Individual candidates could also offer themselves for election, in effect as regional 'lists' containing one name.

The additional member calculation began in each regional area after all of the region's constituency results were declared. The list votes for each party and individual from each constituency in a region were totalled, then divided by the number of constituency seats won by each party and individual within the region, plus one. The party or individual with the highest quotient was given the first regional list seat. To fill the other six seats, this method was used a further six times, with the denominator incremented by one every time a party or individual was allocated a list seat in that region. An example of the additional member system calculation, based on the 1997 General Election results, is contained in Clark and Berridge (1998).

Overview of the Results

Table 5.1 shows the number of votes cast, the percentage share of the vote and the number of Members (MSPs) elected. All four main parties polled fewer votes and achieved a lower vote share in the lists than in the constituencies - particularly Labour. Over 12,000 more votes were cast for list candidates than for constituency candidates, proving that there were some electors who did not support any of the main parties and took advantage of the greater choice offered to them in the lists. The Scottish Socialist Party MSP Tommy Sheridan and the Green MSP Robin Harper owed their election to voters' understanding of how the new system worked, and it is hardly surprising that during the campaign the main parties were desperately trying to convince electors, particularly their own supporters, to give them both a constituency and a list vote. The level of rejected ballots was, on average, three times higher than at the 1997

General Election, and worked out at just over 100 rejected ballots per seat. There was only one constituency - Ayr - where the winning majority (25) was lower than the number of rejected ballots.

Table 5.1   **Votes Cast, Vote Shares and MSPs Elected in 1999**

Party	Constituency votes	Constituency % vote share	Constituency MSPs	List votes	List % vote share	List MSPs	Total MSPs
LAB	896,906	38.4	53	786,818	33.5	3	56
SNP	672,768	28.8	7	638,644	27.2	28	35
CON	359,576	15.4	0	359,109	15.3	18	18
LIBDEM	333,179	14.3	12	290,760	12.4	5	17
GRN	-	-	-	84,023	3.6	1	1
SOCLAB	5,268	0.2	0	55,232	2.4	0	0
SSP	23,654	1.0	0	46,635	2.0	1	1
PLA	-	-	-	9,784	0.4	0	0
SUP	-	-	-	7,011	0.3	0	0
LIB	-	-	-	5,534	0.2	0	0
NLP	-	-	-	4,906	0.2	0	0
HIA	-	-	-	2,607	0.1	0	0
UKIP	-	-	-	1,502	0.1	0	0
SFPP	-	-	-	1,373	0.1	0	0
WTP	-	-	-	1,184	0.1	0	0
CRM	-	-	-	806	0.0	0	0
SPGB	-	-	-	697	0.0	0	0
CPB	190	0.0	0	521	0.0	0	0
HP	-	-	-	447	0.0	0	0
IND	31,901	1.4	1	41,321	1.8	0	1
SWP	2,757	0.1	0	-	-	-	0
Rejected Ballots	7,839	0.3	-	7,268	0.3	-	-
Total	2,334,038	-	73	2,346,182	-	56	129

Source: Based on results from Scottish Executive Justice Department. Minor party codes: GRN = Green; SOCLAB = Socialist Labour Party (5); SSP = Scottish Socialist Party (18); PLA = Pro-Life Alliance; SUP = Scottish Unionist Party; LIB = Liberal Party; NLP = Natural Law Party; HIA = Highlands and Islands Alliance; UKIP = UK Independence Party; SFPP = Scottish Families and Pensioners Party; WTP = Witchery Tour Party; CRM = Civil Rights Movement; SPGB = Socialist Party of Great Britain; CPB = Communist Party of Britain (1); HP = Humanist Party; IND = Independent (18); SWP = Socialist Workers' Party (5). Figures in brackets denote the number of minor party and individual constituency candidates in 1999; major parties contested all seats. Percentages may not sum to 100% due to rounding.

The effect of the new voting system is clearly shown by comparing party vote shares with the number of MSPs elected by each method. Labour won 73% of the constituencies with only 38% of the vote; the Liberal Democrats also won proportionately more seats (16%) than votes (14%). However, the SNP won proportionately fewer seats (10%) than votes (29%), and the Tories again failed to win any constituencies, despite taking 15% of the vote in 1999. In the lists, however, the SNP gained proportionately more seats (50%) than votes (27%), as did the Conservatives (32% of list seats; 15% of list votes). By contrast, Labour won only 5% of the list seats with 34% of the votes, and the Liberal Democrats won 9% of the seats with 12% of the votes. If votes from both elements of the election were added together, Labour had 36% of the vote and 43% of the seats, but the other three parties achieved a much more proportional result (SNP: 28% of the vote, 27% of MSPs; CON: 15% of the vote, 14% of MSPs; LIBDEM: 13% of the vote, 13% of MSPs). So the new voting system merely reduced the imbalance caused by simple plurality, and did not eliminate it entirely.

Table 5.2 compares the 1999 constituency results with those from the 1997 General Election. Labour polled 386,444 fewer votes in 1999, won three fewer seats and suffered a 7.1% reduction in their vote share. The SNP gained over 50,000 more votes and one more seat, while their vote share rose by 6.7% giving, on average, a 6.9% swing from Labour to SNP. Even a swing of this magnitude, however, had little effect on the constituency results, as the SNP was only able to capture one marginal seat from Labour (Inverness East, Nairn & Lochaber) - after 1997 they were, on average, 33.7% behind Labour in the 56 Labour-held seats. Interestingly, the Labour-SNP swing in 1999 is easily higher than the average swings in all post-second World War General Elections except 1997, when there was a 10% swing from the Tories to Labour.

Although the Liberal Democrats polled over 32,000 fewer votes in 1999, their vote share rose by 1.3% (an arithmetical effect caused by the lower turnout) and they won Aberdeen South from Labour and both the new seats in Orkney and Shetland, giving them a net gain of two. The Tories again failed to win a single constituency, and their vote share in the constituencies fell to 15.4%. Labour's third lost seat was Falkirk West where Dennis Canavan MP, having failed to be selected as a prospective candidate for Holyrood, stood as an Independent and won convincingly.

Both the Conservatives and the SNP benefited most from the additional member system, but as long-term proponents of proportional representation the Liberal Democrats may have found its effects on their results somewhat ironic. By more than punching their weight in the constituency elections in five of the eight regional areas, the number of regional members they could get elected was limited. Furthermore, one recalls the common mantra regularly deployed by the two main parties during the post-war period that 'a vote for the Liberals is a wasted vote'. In many constituency seats under a first-past-the-post voting system that may have been the case, but when the opportunity arose to make every vote count there did not appear to be a vast hidden reservoir of Liberal Democrat voters. Far from casting their list votes for the Liberal Democrats or, indeed, any of the main parties, many electors chose instead to support minor parties and individual candidates.

Table 5.2 Constituency Votes, Vote Shares and MPs/MSPs, 1997-1999

Party	1997 votes	1997 % vote share	1997 MPs	1999 votes	1999 % vote share	1999 MSPs
LAB	1,283,350	45.5	56	896,906	38.4	53
SNP	621,550	22.1	6	672,768	28.8	7
CON	493,059	17.5	0	359,576	15.4	0
LIBDEM	365,359	13.0	10	333,179	14.3	12
IND	1,536	0.1	0	31,901	1.4	1
SSP	9,740	0.4	0	23,654	1.0	0
SOCLAB	1,792	0.1	0	5,268	0.2	0
SWP	-	-	-	2,757	0.1	0
CPB	-	-	-	190	0.0	0
REF	26,980	1.0	0	-	-	-
PLA	5,172	0.2	0	-	-	-
NLP	1,922	0.1	0	-	-	-
GRN	1,721	0.1	0	-	-	-
UKIP	1,585	0.1	0	-	-	-
BNP	651	0.0	0	-	-	-
LIB	650	0.0	0	-	-	-
SPGB	315	0.0	0	-	-	-
OTHER	1,363	0.0	0	-	-	-
Rejected Ballots	2,593	0.1	-	7,839	0.3	-
Total	2,819,338	-	72	2,334,038	-	73

Source: 1997 data from Clark and Berridge (1998); 1999 data based on results from Scottish Executive Justice Department. Minor party codes: IND = Independent (4); SSP = Scottish Socialist Party (16 – fought as 'Scottish Socialist Alliance in 1997); SOCLAB = Socialist Labour Party (3); SWP = Socialist Workers' Party; CPB = Communist Party of Britain; REF = Referendum Party (67); PLA = Pro-Life Alliance (9); NLP = Natural Law Party (14); GRN = Green (5); UKIP = UK Independence Party (9); BNP = British National Party (3); LIB = Liberal Party (2); SPGB = Socialist Party of Great Britain (2). Figures in brackets denote the number of candidates fielded by minor parties in 1997 (there were 9 OTHER candidates). Percentages may not sum to 100% due to rounding.

The SNP consolidated their position as second party in Scotland (they came second in 51 of the 73 constituency seats), gaining as many MSPs as the Tories and the Liberal Democrats combined. The new voting system rewarded their relative strength in depth throughout Scotland, with 80% of their MSPs elected from the lists. Whilst undoubtedly the SNP performance was successful insofar as they increased their representation almost six-fold, it did fall short of their predictions that they would take more constituency seats, and their average 28% share of the vote did not eclipse their achievement of 30.4% of Scottish votes in the General

Election of October 1974, when the turnout was much higher at 74.8%. The SNP can take much comfort from being the only main party to receive more constituency votes in 1999 than in 1997, despite the lower turnout, and are consequently in a much stronger position to win constituencies from Labour at the next Holyrood elections. The future prognosis for the SNP is quite difficult to determine, as a modest increase in the number of constituencies they win would not necessarily lead to significantly greater representation, due to the way that the list calculation works. In five of the eight regions in 1999, the SNP took the seventh and final list seat (Curtis, 1999), and so under similar circumstances, it is possible that if they won one extra constituency in each region, five of these gains could be cancelled out by winning fewer list seats. Much may depend on whether the turnout at future Holyrood elections can reach the levels normally seen at UK General Elections and whether the share of a much larger vote would be distributed as, or more, favourably to the SNP.

The Conservatives must have approached the Holyrood election with some ambivalence. On the one hand, they were seeking election to a body that many Tories had actively campaigned against in the 1997 referendum. More crucially, perhaps, for their long-term prospects as a political force, the new voting system seemed likely to give them an opportunity to become a major player in Scottish politics again, after their total wipe-out in the 1997 General Election. In the sense that it was the first step out of the wilderness, their performance could be regarded as a success - after all, anything is surely better than nothing - but in failing to win any constituencies for the second successive election, it is clear that to many voters, the Scottish Tories are still too closely linked with the Conservative Party which fell from power in such a spectacular fashion in 1997. It must have been a source of concern that their vote share fell again, and perhaps the best spin that could be applied to this fact is that many former Conservative voters had opposed devolution and therefore declined to vote for a Parliament they had not wanted. (Average turnout figures do not wholly support this contention - see the following section.) If that hypothesis is correct, however, in order to do better in future Holyrood elections they will have to both persuade their former voters to support them and, ideally, attract new ones. As to their performance in the next Westminster election, it seems likely that they face a similar and daunting task if recent public opinion poll ratings are to be believed. All the signs indicate it could be a long road back for the party that in 1955 secured over 50% of the vote and over 50% of MPs in Scotland.

Labour, as the party that delivered devolution, might have hoped to be rewarded with a clear majority of MSPs. The reality was somewhat different. Compared with the 1997 General Election, they received over

386,000 fewer votes in the constituencies and over 110,000 fewer votes again in the lists than in the constituencies. Nonetheless, their share of the vote was still nearly 10% higher than the SNP's, and by retaining all but three of their constituencies they were guaranteed to be the largest party, though nine seats short of an overall majority. Labour's performance in Scotland in 1997 was unusually high - for example, 39 of their 56 winning candidates received over 50% of the vote. Given the size of the 1997 swing, it is probable that they picked up many votes from disaffected former Tory supporters and also tactical votes from supporters of other parties to ensure that Conservatives were defeated, but even their biggest optimist would have found it difficult to expect Labour to do as well as that again. In fact, their 1999 constituency vote share was close to the 39% they received in Scotland at the 1992 General Election, suggesting that their current level of support is more realistic, although given the lower turnout it represents a similar share of a much smaller pool of votes. Another possible reason for the slump in Labour's support is that the 1999 election coincided with the mid-term period of the Blair Government at Westminster when incumbent governments historically have sustained a drop in support. Although still perceived by public opinion polls to be popular (and more popular than most mid-term governments in recent times) Labour could reasonably expect a fall in their support at this point in the UK electoral cycle. While in time the effectiveness of the Scottish Parliament is likely to be judged on its own merits, rather on what has happened at Westminster, it may be that Labour as a government was judged by some electors and found wanting.

High and Low Turnouts

The average Scottish turnout of 58% masked considerable variations between individual constituencies and, to a lesser extent, regional areas. The turnout for list votes was normally within 0.3% of the constituency turnout for each seat. The highest turnouts in constituency seats were Stirling (67.2%), Eastwood and Edinburgh West (both 66.8%), Strathkelvin & Bearsden (66.5%) and Galloway & Upper Nithsdale and Ayr (both 65.9%). All were formerly Tory-held constituencies, in which turnouts have been historically higher. The lowest constituency turnouts were all in Labour-held Glasgow seats: Shettleston (40.6%), Maryhill (40.8%), Springburn (43.6%), Kelvin (46.3%), Baillieston (48.0%) and Govan (49.5%). Comparing both groups, the reduction in turnout since 1997 was between 11.4% and 17.3% in the high-turnout seats and between 10.4% and 15.7% in the low-turnout seats - not significantly different, and fitting the average reduction in turnout throughout Scotland quite well (13.8%). Within regional list areas, the turnout for list votes varied between 61.9% in Scotland South and 48.0% in Glasgow. The only other region whose turnout was lower than the average was North-East Scotland (54.7%).

Whilst it is likely that abstentions by voters who opposed the establishment of a Scottish Parliament were partly responsible for the reduction in turnout, this is unlikely to be the only, or even the main, reason. Reversing the argument, in many constituencies a sizeable proportion of the electorate supported parties who were in favour of devolution, but this factor alone did not guarantee a high turnout, or even as high a turnout as in the 1997 referendum (just over 60%). Furthermore, a majority of the sitting MPs declined to stand for election to Holyrood, as did most of those defeated in 1997, and so it is fair to say that many of the candidates were not well known to electors, other than ardent watchers of local and national politics. If the electorate sees legislation enacted by MSPs and their work in a positive light, the turnout could well rise at the next Holyrood election in 2003, but any disillusionment could lead to lower turnouts, as are often found in local and European elections.

Highest and Lowest Majorities in Constituency Seats

Interestingly, the Labour, Liberal Democrat and SNP MSPs with the highest majorities were the party leaders: Jim Wallace (Lib Dem; 51.6% majority in Orkney); First Minister Donald Dewar (Lab; 38.5% in Glasgow Anniesland) and Alex Salmond (SNP; 35.5% in Banff & Buchan). Dennis Canavan (Independent) achieved a 36.1% majority. Of the four, only Dewar failed to increase his majority, which was down 6.2% on 1997. The others' majorities rose by 0.2% (Canavan), 3.6% (Salmond) and 17.9% (Wallace, in the new Orkney seat).

After the 1997 General Election there were only nine marginal seats (where the winner's majority was 10% of votes cast or less). Labour held four, the Liberal Democrats held three and the SNP held two. By 1999 the number of marginals had more than doubled, with 13 Labour seats and three each held by the SNP and the Liberal Democrats. Two of the Labour marginals - Aberdeen North (majority 1.4%) and Dundee West (0.4%) - were very nearly disasters for the party, as they had unquestionably been regarded as safe seats in 1997. The SNP nearly captured both seats, in tandem with the significant gains they made in the local government elections that were held on the same day.

Westminster MPs Standing for Holyrood

In all, 15 of the 72 Scottish MPs contested the Holyrood election and 14 won a constituency seat (the Liberal Democrat MP for Edinburgh West, Donald Gorrie, did not stand in a constituency but won a list seat in Central Scotland). Apart from Wallace, Canavan and Salmond the only other MSPs who increased their majorities since 1997 were Andrew Welsh (SNP; +2.1% in Angus) and John Swinney (SNP; +1.9% in North Tayside). The other nine all lost some ground, particularly the six Labour MSPs, though by noticeably less than other candidates defending seats Labour had won in 1997. Only Wallace and Welsh increased their vote shares (by 15.2% and 1.8% respectively), and of the other incumbent MPs, only one sustained a drop in vote share of more than 5% (John McAllion; Labour, Dundee East; -7.9%).

As incumbents defending seats already held, the SNP MPs did comparatively worse than their colleagues, because there was little scope for them to add substantially to their vote. By contrast, most of the SNP challengers increased their vote shares due to the large falls in turnout and Labour support. However, there were other seats, notably Glasgow Govan, which the SNP had almost certainly expected to gain but failed to do so. In Govan (one of the four Labour-held marginals after 1997), more than in any of the other Glasgow seats, the Labour vote share held firm - in fact there was very little change in the distribution of votes at all from the 1997 result. And in Ochil (technically not a marginal, but within range of the kind of swings which ousted the Tories in 1997) there was a similar outcome with Labour doing comparatively better and the SNP comparatively worse than in many of the other neighbouring Labour-held seats.

Analysis at the Regional Level

Finally, Tables 5.3 to 5.10 show votes, vote shares and MSPs elected to both constituency and list seats in all eight regions.

If Dennis Canavan had failed to win Falkirk West, the level of support he drew from the surrounding constituencies in Central Scotland (Table 5.3) would easily have given him a list seat (all but 248 of the votes for Independents were for him). Compared with the 1997 results, Labour's constituency vote share fell by 12.9% (5.6% of that transferred to Canavan as an Independent), as did the Conservatives' (down 0.8%). The SNP and Liberal Democrat vote shares rose, by 6.4% and 1.4% respectively. There were only two marginal constituencies - Ochil (majority 3.5%) and Kilmarnock & Loudoun (majority 7.0%); both were Labour seats.

Table 5.3 Central Scotland

Party	Constituency votes	Constituency % vote share	Constituency MSPs	List votes	List % vote share	List MSPs	Total MSPs
LAB	154,065	46.4	9	129,822	39.2	0	9
SNP	98,858	29.8	0	91,802	27.7	5	5
CON	31,767	9.6	0	30,243	9.1	1	1
IND	18,869	5.7	1	27,948	8.4	0	1
LIBDEM	21,911	6.6	0	20,505	6.2	1	1
SOCLAB	4,648	1.4	0	10,956	3.3	0	0
GRN	-	-	-	5,926	1.8	0	0
SSP	1,116	0.3	0	5,739	1.7	0	0
SUP	-	-	-	2,888	0.9	0	0
PLA	-	-	-	2,567	0.8	0	0
SFPP	-	-	-	1,373	0.4	0	0
NLP	-	-	-	719	0.2	0	0
Rejected Ballots	1,109	0.3	-	1,094	0.3	-	-
Total	332,343	-	10	331,582	-	7	17

Source: Based on results from Scottish Executive Justice Department. Minor party codes: IND = Independent (2); SOCLAB = Socialist Labour Party (3); GRN = Green; SSP = Scottish Socialist Party (1); SUP = Scottish Unionist Party; PLA = Pro-Life Alliance; SFPP = Scottish Families and Pensioners Party; NLP = Natural Law Party. Figures in brackets denote the number of minor party and individual constituency candidates in 1999; major parties contested all seats. Percentages may not sum to 100% due to rounding.

It is noticeable that there were 10,694 more list votes than constituency votes in Glasgow (Table 5.4), clearly part of a concerted (and successful) effort to elect the Scottish Socialist, Tommy Sheridan. The Liberal Democrats were third party in terms of constituency votes, but only fifth party in terms of list votes. Comparing the constituencies with the 1997 result, Labour's vote share fell by 13.5%, while the SNP vote share rose by 9.8%, giving a much higher swing than the Scottish average, but the SNP would have to nearly double their vote share at the next election to trouble Labour in what is still Labour's strongest region, with only the Govan seat (majority 6.6%) classed as a marginal. The Tory vote share also fell slightly, by 0.4%, but the Liberal Democrat vote share rose by 1.1%.

Table 5.4 Glasgow

Party	Constituency votes	Constituency % vote share	Constituency MSPs	List votes	List % vote share	List MSPs	Total MSPs
LAB	115,344	46.8	10	112,588	43.8	0	10
SNP	71,566	29.0	0	65,360	25.4	4	4
CON	19,897	8.1	0	20,239	7.9	1	1
SSP	16,177	6.6	0	18,581	7.2	1	1
LIBDEM	20,756	8.4	0	18,473	7.2	1	1
GRN	-	-	-	10,159	4.0	0	0
SOCLAB	620	0.3	0	4,391	1.7	0	0
PLA	-	-	-	2,357	0.9	0	0
SUP	-	-	-	2,283	0.9	0	0
CPB	190	0.1	0	521	0.2	0	0
HP	-	-	-	447	0.2	0	0
NLP	-	-	-	419	0.2	0	0
SPGB	-	-	-	309	0.1	0	0
IND	-	-	-	221	0.1	0	0
SWP	920	0.4	0	-	-	-	0
Rejected Ballots	1,169	0.5	-	985	0.4	-	-
Total	246,639	-	10	257,333	-	7	17

Source: Based on results from Scottish Executive Justice Department. Minor party codes: SSP = Scottish Socialist Party (9); GRN = Green; SOCLAB = Socialist Labour Party (2); PLA = Pro-Life Alliance; SUP = Scottish Unionist Party; CPB = Communist Party of Britain (1); HP = Humanist Party; NLP = Natural Law Party; SPGB = Socialist Party of Great Britain; IND = Independent; SWP = Socialist Workers' Party (1). Figures in brackets denote the number of minor party and individual constituency candidates in 1999; major parties contested all seats. Percentages may not sum to 100% due to rounding.

The political makeup of Highlands and Islands is considerably different from that of the Central belt. The Liberal Democrats took five constituency seats, although they received fewer votes than the SNP, who won two constituencies. The list seats were shared by Labour, the SNP and the Tories in the ratio 3:2:2. There was little change in constituency vote shares compared with 1997: Labour up by 0.3%, SNP up by 1.7%, the Liberal Democrats up by 0.5% and the Tories down by 1.9%. There were two marginal seats: Argyll & Bute (Libdem majority 6.4%) and Inverness East, Nairn & Lochaber (SNP majority 1.1%).

Table 5.5 Highlands and Islands

Party	Constituency votes	Constituency % vote share	Constituency MSPs	List votes	List % vote share	List MSPs	Total MSPs
SNP	57,555	28.4	2	55,933	27.6	2	4
LAB	55,236	27.3	1	51,371	25.3	3	4
LIBDEM	57,034	28.1	5	43,226	21.3	0	5
CON	28,890	14.3	0	30,122	14.9	2	2
GRN	-	-	-	7,560	3.7	0	0
IND	3,138	1.6	0	5,739	2.8	0	0
SOCLAB	-	-	-	2,808	1.4	0	0
HIA	-	-	-	2,607	1.3	0	0
SSP	-	-	-	1,770	0.9	0	0
NLP	-	-	-	536	0.3	0	0
Rejected Ballots	805	0.4	-	1,034	0.5	-	-
Total	202,658	-	8	202,706	-	7	15

Source: Based on results from Scottish Executive Justice Department. Minor party codes: GRN = Green; IND = Independent (3); SOCLAB = Socialist Labour Party; HIA = Highlands and Islands Alliance; SSP = Scottish Socialist Party; NLP = Natural Law Party. Figures in brackets denote the number of minor party and individual constituency candidates in 1999; major parties contested all seats. Percentages may not sum to 100% due to rounding.

Lothians (Table 5.6) is another central belt area dominated by Labour-held constituencies. Perhaps for that reason, their vote share dropped by a quarter in the lists compared with the constituencies, as Labour voters realised the difficulty of electing any additional Members. The main beneficiary of this was undoubtedly Robin Harper, who won the Green Party's first seat in a national election in Scotland. Compared with 1997, the Labour and Conservative constituency vote shares were down by 5.8% and 3.4% respectively, while the SNP and Liberal Democrat vote shares rose by 8.4% and 0.7% respectively. There were two Labour-held marginals: Edinburgh Pentlands (majority 7.3%) and Linlithgow (majority 8.6%).

Table 5.6 Lothians

Party	Constituency votes	Constituency % vote share	Constituency MSPs	List votes	List % vote share	List MSPs	Total MSPs
LAB	133,612	40.1	8	99,908	30.2	0	8
SNP	89,306	26.8	0	85,085	25.7	3	3
CON	52,778	15.8	0	52,067	15.7	2	2
LIBDEM	52,138	15.6	1	47,565	14.4	1	2
GRN	-	-	-	22,848	6.9	1	1
SOCLAB	-	-	-	10,895	3.3	0	0
SSP	2,434	0.7	0	5,237	1.6	0	0
LIB	-	-	-	2,056	0.6	0	0
WTP	-	-	-	1,184	0.4	0	0
IND	1,722	0.5	0	1,014	0.3	0	0
PLA	-	-	-	898	0.3	0	0
CRM	-	-	-	806	0.2	0	0
NLP	-	-	-	564	0.2	0	0
SPGB	-	-	-	388	0.1	0	0
SWP	482	0.2	0	-	-	-	0
Rejected Ballots	1,012	0.3	-	912	0.3	-	-
Total	333,484	-	9	331,427	-	7	16

Source: Based on results from Scottish Executive Justice Department. Minor party codes: GRN = Green; SOCLAB = Socialist Labour Party; SSP = Scottish Socialist Party (3); LIB = Liberal Party; WTP = Witchery Tour Party; IND = Independent (5); PLA = Pro-Life Alliance; CRM = Civil Rights Movement; NLP = Natural Law Party; SPGB = Socialist Party of Great Britain; SWP = Socialist Workers' Party (1). Figures in brackets denote the number of minor party and individual constituency candidates in 1999; major parties contested all seats. Percentages may not sum to 100% due to rounding.

In Mid Scotland and Fife (Table 5.7) there were 4,780 more list votes than constituency votes, with the Conservatives, as opposed to the minor parties, being the main beneficiaries. The region contains three seats which the Tories held until 1997, and is one of only three regions in which they elected more than two MSPs (the others being North East Scotland and Scotland South). Compared with 1997, the Labour and Conservative constituency vote shares were down by 3.1% and 3.9% respectively, while the SNP and Liberal Democrat vote shares rose by 6.6% and 0.4% respectively. Only Perth was marginal (SNP majority 5.4%).

Table 5.7 Mid Scotland and Fife

Party	Constituency votes	Constituency % vote share	Constituency MSPs	List votes	List % vote share	List MSPs	Total MSPs
LAB	111,154	36.9	6	101,964	33.3	0	6
SNP	96,102	31.9	2	87,659	28.6	3	5
CON	51,860	17.2	0	56,719	18.5	3	3
LIBDEM	39,191	13.0	1	38,896	12.7	1	2
GRN	-	-	-	11,821	3.9	0	0
SOCLAB	-	-	-	4,266	1.4	0	0
SSP	-	-	-	3,044	1.0	0	0
PLA	-	-	-	735	0.2	0	0
NLP	-	-	-	558	0.2	0	0
IND	2,432	0.8	0	-	-	-	0
Rejected Ballots	805	0.3	-	662	0.2	-	-
Total	301,544	-	9	306,324	-	7	16

Source: Based on results from Scottish Executive Justice Department. Minor party codes: GRN = Green; SOCLAB = Socialist Labour Party; SSP = Scottish Socialist Party; PLA = Pro-Life Alliance; NLP = Natural Law Party; IND = Independent (3). Figures in brackets denote the number of minor party and individual constituency candidates in 1999; major parties contested all seats. Percentages may not sum to 100% due to rounding.

North East Scotland was one of the most closely contested regions, with the constituency seats fairly evenly shared between Labour, the Liberal Democrats and the SNP. Table 5.8 illustrates the inequity of the first past the post system very well: the SNP took nearly 7% more of the vote throughout the region, but only half as many constituencies as Labour. However, this could change in future, as a drop in Labour support of nearly 40,000 votes allowed the SNP, who held position, to overtake them. Furthermore, five of the nine constituencies are now marginal: three were Labour-held with the SNP challenging (Aberdeen North, majority 1.4%; Dundee West, majority 0.4% and Dundee East, majority 9.0%). The Liberal Democrats held the other two marginals (Aberdeen South, majority 5.1% and West Aberdeenshire and Kincardine, majority 6.4%). Compared with 1997, the Labour and Conservative constituency vote shares declined by 4.7% and 4.6% respectively, and the SNP and Liberal Democrat vote shares rose by 6.9% and 2.3% respectively.

Table 5.8 North East Scotland

Party	Constituency votes	Constituency % vote share	Constituency MSPs	List votes	List % vote share	List MSPs	Total MSPs
SNP	94,462	33.0	2	92,329	32.3	4	6
LAB	74,648	26.1	4	72,666	25.4	0	4
CON	50,901	17.8	0	52,149	18.2	3	3
LIBDEM	60,540	21.2	3	49,843	17.4	0	3
GRN	-	-	-	8,067	2.8	0	0
SOCLAB	-	-	-	3,557	1.2	0	0
IND	2,559	0.9	0	3,073	1.1	0	0
SSP	2,063	0.7	0	3,016	1.1	0	0
NLP	-	-	-	746	0.3	0	0
SWP	206	0.1	0	-	-	-	0
Rejected Ballots	807	0.3	-	750	0.3	-	-
Total	286,186	-	9	286,196	-	7	16

Source: Based on results from Scottish Executive Justice Department. Minor party codes: GRN = Green; SOCLAB = Socialist Labour Party; IND = Independent (1); SSP = Scottish Socialist Party (3); NLP = Natural Law Party; SWP = Socialist Workers' Party (1). Figures in brackets denote the number of minor party and individual constituency candidates in 1999; major parties contested all seats. Percentages may not sum to 100% due to rounding.

Scotland South (Table 5.9) was the only region in 1999 in which the Conservatives received more than 20% of the constituency votes.

Table 5.9 Scotland South

Party	Constituency votes	Constituency % vote share	Constituency MSPs	List votes	List % vote share	List MSPs	Total MSPs
LAB	117,799	36.8	6	98,836	30.9	0	6
SNP	81,516	25.5	1	80,059	25.1	3	4
CON	72,690	22.7	0	68,904	21.6	4	4
LIBDEM	46,639	14.6	2	38,157	12.0	0	2
SOCLAB	-	-	-	13,887	4.4	0	0
GRN	-	-	-	9,468	3.0	0	0
LIB	-	-	-	3,478	1.1	0	0
SSP	-	-	-	3,304	1.0	0	0
UKIP	-	-	-	1,502	0.5	0	0
NLP	-	-	-	775	0.2	0	0
Rejected Ballots	1,108	0.4	-	1,029	0.3	-	-
Total	319,752	-	9	319,399	-	7	16

Source: Based on results from Scottish Executive Justice Department. Minor party codes: SOCLAB = Socialist Labour Party; GRN = Green; LIB = Liberal Party; SSP = Scottish Socialist Party; UKIP = UK Independence Party; NLP = Natural Law Party. Major parties contested all seats. Percentages may not sum to 100% due to rounding.

Though they failed to recapture Ayr by only 25 votes, they did take four of the seven list seats. Labour held two thirds of the constituencies, but three of their six seats are now marginal: Clydesdale (majority 9.9%), Dumfries (majority 9.5%) and Ayr (majority 0.1%, with a by-election looming). The SNP's only constituency in this region is also marginal (Galloway & Upper Nithsdale, majority 9.0%). Of the main parties, only Labour sustained a drop in constituency vote share compared with 1997 (by 6.5%). The vote shares of the SNP, the Tories and the Liberal Democrats rose by 6.5%, 0.1% and 1.2% respectively.

Table 5.10 West of Scotland

Party	Constituency votes	Constituency % vote share	Constituency MSPs	List votes	List % vote share	List MSPs	Total MSPs
LAB	135,048	43.4	9	119,663	38.5	0	9
SNP	83,403	26.8	0	80,417	25.8	4	4
CON	50,793	16.3	0	48,666	15.6	2	2
LIBDEM	34,970	11.2	0	34,095	11.0	1	1
GRN	-	-	-	8,174	2.6	0	0
SSP	1,864	0.6	0	5,944	1.9	0	0
SOCLAB	-	-	-	4,472	1.4	0	0
IND	3,181	1.0	0	3,326	1.1	0	0
PLA	-	-	-	3,227	1.0	0	0
SUP	-	-	-	1,840	0.6	0	0
NLP	-	-	-	589	0.2	0	0
SWP	1,149	0.4	0	-	-	-	0
Rejected Ballots	1,024	0.3	-	802	0.3	-	-
Total	311,432	-	9	311,215	-	7	16

Source: Based on results from Scottish Executive Justice Department. Minor party codes: GRN = Green; SSP = Scottish Socialist Party (2); SOCLAB = Socialist Labour Party; IND = Independent (4); PLA = Pro-Life Alliance; SUP = Scottish Unionist Party; NLP = Natural Law Party; SWP = Socialist Workers' Party (2). Figures in brackets denote the number of minor party and individual constituency candidates in 1999; major parties contested all seats. Percentages may not sum to 100% due to rounding.

The West of Scotland region gave Labour their third-best constituency vote share in 1999 (Table 5.10). Labour won all the constituencies, as they did in Glasgow, including two marginals: Eastwood (majority 4.7%) and West Renfrewshire (majority 8.5%). The Labour and Conservative constituency vote shares were down on 1997 by 7.9% and 1.9% respectively. The SNP and Liberal Democrat vote shares rose by 6.9% and 2.0% respectively.

References

Clark, I. and Berridge J. (1998), *Scotland Votes: The General Election 1997 in Scotland*, University of Dundee General Election Studies.

Curtis, S. (1999), *Order of Election of MSPs from Regional Vote*, The Scottish Parliament Information Centre, Research Note 4/1999.

6 Small Parties in a Devolved Scotland

LYNN G. BENNIE

Introduction

Historically, new and small parties in Scotland have found it difficult to make an impression on what has essentially been a four-party system.[1] However, the creation of a Scottish Parliament and the adoption of an element of proportionality in the electoral system used to elect MSPs has led to discussion of a changing party system in Scotland. In the run-up to the 1999 elections, there were predictions of a more diverse party system, and of new, different voices being represented in the Parliament. In the event, the Scottish Socialist Party (SSP) and the Scottish Green Party (SGP) successfully entered the Parliament, with one representative each, and Dennis Canavan was elected as an independent, having failed to be selected as a candidate for Labour.

This chapter examines changes in Scotland's party system, with an emphasis on the position of the small parties.[2] It considers the performance of these parties in British General Elections and examines their election experiences in 1999. Finally, the chapter assesses the potential influence of those small parties that enjoyed electoral success in May 1999.[3] Müller-Rommel and Pridham (1989) outline a number of ways of measuring party relevance, starting with numerical variables (number of votes and so on), progressing to consideration of political variables (political function and legitimacy). In a similar fashion, we can assess the small parties in the new Scottish political arena by examining the number of parties contesting elections; the number of small party candidates; the number (and percentage) of votes these parties attract; the number of seats won; and the role of these parties when they enter the legislature.

Small Parties in British General Elections

Small parties have always existed in Britain, but on the fringes of politics. The British political system has discouraged the growth of small parties; in particular, the first-past-the-post electoral system has proved an almost insurmountable barrier. Between 1959 and 1992, in England, Scotland and Wales, only four candidates other than major party candidates achieved representation at Westminster - two independent Labour candidates, and two independent Conservatives (Butler and Butler 1994: 157-158). Small party candidates have contested British elections *expecting* to lose their election deposits: The Communist Party of Great Britain put forward 571 candidates between 1922 and 1992, and lost 534 of them; the Green Party and its predecessors (People/ Ecology Party) put forward 589 candidates between 1974 and 1993, losing all but one deposit (Butler and Butler 1994: 157-158).[4] So, in terms of parties able to gain representation, Britain's party system has been far from multi-party in character. Even at the local level, where there is more of a tradition of non-partisan politics, there has been an increasing dominance of the major parties, with an increase in the number of major party candidates, less independent candidates, and increasing competition and partisanship (Denver and Bochel 1994).

In Scotland between 1974 and 1997 the small parties and independents never achieved more than 2% of the vote in General Elections, and their performance was related to that of the main opposition parties. For example, when the SNP was doing well in the 1970s, small parties and independent candidates benefited less from protest voting: in October 1974 non-major party candidates attracted only 7,721 votes (0.28%). Likewise, at the time of the 1983 General Election, when the SDP/ Liberal Alliance was at its height, 'others' attracted only 7,830 votes (also 0.28%). In 1997, the 'other' candidates did relatively well by their own standards, with 1.91% of the vote (53,408 votes), more than double the number of votes gained in 1992. In 1997 there was a larger than usual number of parties standing, along with a larger than usual number of small party and independent candidates, largely due to Goldsmith's Referendum Party, which fielded 67 candidates. No other small party fielded more than 20 candidates.

In 1997 the Scottish Socialist Alliance (SSA) fielded 16 candidates and performed relatively well in a number of Glasgow constituencies.[5] In Glasgow Ballieston the SSA came fifth behind the major parties and ahead of the Referendum Party with 970 votes (3.05%). The party's best result was in Glasgow Pollok where Tommy Sheridan came third ahead of the Conservative and Liberal Democrat candidates (with nearly double the Conservative vote and nearly three times that of the Liberal Democrats). Scargill's Socialist Labour Party (SLP) stood in only three constituencies, its best performance in Motherwell and Wishaw, with 797 votes (2.18%). The Greens meanwhile were virtually non-existent in these elections; they stood in only five constituencies and attracted an average vote of 0.84%.

The party's best result was in Edinburgh Central, with 607 votes for the candidate Linda Hendry, only 1.42%.

An interesting feature of 1997 was the rise of 'pressure parties', namely the Referendum Party, the Pro Life Alliance, and the UK Independence Party, highlighting the complex relationship between pressure groups and small political parties. Goldsmith's Referendum Party was regarded as a 'one-man band', with little formal organisation, but it fielded more than 500 candidates throughout the UK in 1997 and spent an estimated £20M (*The Guardian* 3 March 1997). In Scotland the party's best result was in Orkney & Shetland where it attracted 820 votes (3.97%). The Pro Life Alliance attracted more than 2% of the vote, and finished fifth, in three out of nine contested constituencies - East Kilbride, Hamilton South and Glasgow Cathcart. In Glasgow Cathcart, the party finished ahead of both the SSA and the Referendum Party. The UK Independence Party also had nine candidates in Scotland; it achieved its best result in Caithness, Sutherland and Easter Ross where it attracted 0.73% of the vote (212 votes).

It is clear, in the context of British elections, small parties and independent candidates have not tended to influence election results, never mind the dynamics of legislatures. The dominant opinion is that these parties have been largely irrelevant. Patrick Dunleavy commented on the rise of small (pressure) parties during the 1997 Westminster elections: 'They (Referendum Party, Pro Life Alliance Party, UK Independence Party) will get no recognition and everybody will make fun of them. Then they will lapse again' (*The Independent* 7 February 1997). Dunleavy used a Downsian argument to explain the development of single-issue groups who stand in elections to promote their cause, in that the main parties tend to adjust to these demands and thus negate the small parties' reasons for standing. However, it could be argued that these parties had an impact on politics more broadly defined. In 1997, the so-called wave of fringe candidates forced the major parties to answer questions on abortion, gun control, the transportation of live animals and so on. They took part in hustings and debates, and got television coverage, and the issues they promoted were discussed during the election.

The Scottish Parliamentary Elections

At the time of the creation of the Scottish Parliament there was talk of increased 'democratic space' with more opportunities for minority interests. There is little doubt that the element of proportional representation in the new electoral system raised the expectations of existing small parties, as well as encouraging new parties to form. The small parties with generally low levels of evenly distributed support had opportunities for election through the regional party lists. Furthermore, the rules governing elections in 1999 were significantly more encouraging for small parties than in General Elections. A £500 deposit was required to stand in a regional list which could include up to 12 named candidates. It therefore cost a party £4,000 to field a list of candidates in each region and, technically speaking, this could pay for between 8 and 96 candidates. Such a figure compares to £36,000 (72 x £500) which is the cost of fielding a full slate of candidates in a General Election.[6]

An examination of the number of small parties standing in the regions in May 1999 reveals that approximately 25 different small parties stood (plus independents), more than twice as many as in the General Election of 1997.[7] The Lothians region was particularly busy with small parties and fringe candidates; this area attracted a total of 17 lists.[8] However, the only small parties to stand in all the regions were the Scottish Greens, the Scottish Socialist Party, the Socialist Labour Party and the Natural Law Party.

The first-past-the-post constituencies were much less attractive to the small parties. In all, less than 50 candidates did not belong to one of the four main parties: 18 Scottish Socialist Party; five Socialist Workers Party; four Socialist Labour Party, one Communist Party and another 20 or so independents. Indeed, less than half of the constituencies - 35 - had a candidate from a small party or an independent. Apart from independents, only the Socialist Workers Party (SWP) stood in the constituencies but not the regions, with five candidates - in Aberdeen South, Edinburgh South, Glasgow Cathcart, Glasgow Rutherglen, and Renfrewshire West. Table 6.1 illustrates the concentration of small party efforts on the lists. Of all the small parties in the constituencies, the SSP clearly dominated. Three SSP candidates finished third, ahead of the Conservative and Liberal Democrat candidates: In Glasgow Pollok Tommy Sheridan polled 5,611 votes (21.5%); in Glasgow Shettleston Rosie Kane came third with 1,640 votes (8%); and in Glasgow Ballieston James McVicar attracted 1,864 votes (7.86%). The highest number of votes to go to an independent, other than Dennis Canavan, was in Gordon where Hamish Watt received 2,559 votes (7.6%) and finished fifth.

Table 6.1 Small Party Candidates 1999

Party	N of regions	Regional Constituency candidates	candidates
Scottish Green Party	8	41	0
Scottish Socialist Party	8	55	18
Highlands and Islands	1	10	0
Socialist Labour Party	8	57	4
Natural Law Party	8	50	0
Pro Life Alliance	5	33	0
Socialist Party GB	2	14	0
Scottish Unionist Party	3	5	0
Humanist Party	1	10	0
Civil Rights Movement	1	8	0
Liberal Party	1	2	0
UK Independence Party	1	1	0
Communist Party GB	1	1	1
Socialist Workers Party	0	0	5

Table 6.2 1997 General Election and 1999 Scottish Parliamentary Results: SSP, Green and Others

Scottish Socialist Party

	Votes	% Votes
1997	9,740	0.35
1999 Constituency Vote	23,654	1.01
1999 Regional Vote	46,635	1.99

Scottish Green Party

	Votes	% Votes
1997	1, 721	0.06
1999 Constituency Vote	0	0
1999 Regional Vote	84, 024	3.59

MP for Falkirk West Dennis Canavan

	Votes	% Votes
1997	0	0
1999 Constituency Vote	18,511	0.79
1999 Regional Vote	27,700	1.18

Others

	Votes	% Votes
1997	41,947	1.49
1999 Constituency Vote	21,662	0.93
1999 Regional Vote	105,221	4.50

Source: Adapted from the Scottish Politics Web Pages (http://www.alba.org.uk).

Even more impressive is the rise in votes for the small parties in 1999. Overall, the small parties and independents attracted just over 11% of the regional vote, and just under 3% of the constituency vote. The small party success story of these elections was, of course, the election of one SSP and one Green candidate. Sheridan was elected in Glasgow with 7.25% (18,581 votes); Harper was elected in the Lothians region with 6.91% of the vote (22,848 votes). The biggest 'other' party was the Socialist Labour Party which won over 50,000 regional votes (just over 2%), more in fact than the SSP. Table 6.2 shows the rise in the number of votes for small parties between 1997 and 1999. The small party disappointment of these elections was the poor performance of the Highlands and Islands Alliance; with only 1.29% (2,607 votes) the Alliance failed to make an impact.

European Parliamentary Elections in Scotland

The June 1999 European Parliamentary Elections were contested for the first time in Britain under a system of proportional representation.[9] Although the number of parties standing in these elections didn't increase between 1994 and 1999, the number of votes increased considerably (see Table 6.3). The underlying trend has been one of increasing support for the small parties but this was certainly exacerbated in 1999.[10] In 1979, the first European Election, no small party took part in Scotland. By 1999 the non-major parties had achieved nearly 15% of the total vote.

The Greens were the only prominent small party in European Elections until 1999. The Greens had experienced a short-term rise in support in 1989, a pattern that was even more pronounced in other parts of the UK (see Rüdig et al 1996).[11] In 1989 Robin Harper attracted 10.47% of the vote (22,331 votes) in Lothians, and the Greens finished ahead of the newly formed Liberal Democrats. However, by 1999 the Greens were competing with the SSP for the position of fifth party. As the table indicates, the Greens maintained fifth place with 5.78%, a rise of 4.18% on their 1994 result. The SSP also improved on Militant's 1994 performance, attracting just over 4%, a gain of 3.22%.

Table 6.3 European Election Performance of Small Parties in Scotland

1984	Party	Votes	%	Candidates
	Ecology	2,560	0.2	1
	Total	2,560	0.2	1
1989	Party	Votes	%	Candidates
	Green	115, 028	7.2	8
	Communist	1,164	0.07	1
	International Communist	193	0.01	1
	Total	116,385	7.28	10
1994	Party	Votes	%	Candidates
	Green	23,304	1.6	8
	Militant Lab	12,113	0.8	1
	Natural Law	5,037	0.3	8
	Liberal	3,249	0.2	1
	Socialist	1,832	0.1	2
	UK Indep	1,096	0.1	1
	Communist	689	0.05	1
	N-E Ethnic	584	0.04	1
	International Communist	381	0.03	1
	Total	48,285	3.22	24
1999	Party	Votes	%	Candidates
	Green	57, 142	5.78	-
	SSP	39, 720	4.02	-
	Pro Euro Cons	17, 781	1.80	-
	UK Indep	12, 459	1.26	-
	Socialist Lab	9,385	0.95	-
	BNP	3,729	0.38	-
	Natural Law	2,087	0.21	-
	Accountant for Lower Scottish Taxes	1,632	0.17	-
	Total	143,935	14.57	-

Campaign Experiences

The Scottish Socialist Party

Scottish Militant Labour had been formed in 1992 when a group of Militant members decided to formally abandon 'entryism' and work independently of the Labour Party. Tommy Sheridan came to prominence as president of the Anti-Poll Tax Federation when he was imprisoned for six months for fighting a warrant sale. His mother, Alice, ran his election campaign and he was elected as a Militant Labour councillor for Pollok in 1992. Sheridan's success came alongside the election of several other Scottish Militant district and regional councillors in 1992, a fact that is often overlooked. In the 1994 European Elections, Sheridan attracted 7.6% across Glasgow. By 1997, the Scottish Socialist Alliance, a collection of left-wing groupings and parties including Scottish Militant Labour was formed to fight the General Election, and the SSP was formed shortly after, in 1998.[12] The new party was a considerable achievement as it formally brought together former Militant Labour supporters, ex-Communists and community and environmental activists. The party was launched by Sheridan with the backing of the Euro MP Hugh Kerr, who was expelled from the Labour Party for criticising the 'right-wing' approach of Tony Blair.[13]

The SSP fought just over 100 local council seats on 6th May 1999, as well as all regional list seats and 18 constituencies. At the launch of the SSP manifesto, Sheridan suggested the party's policies would appeal to the distinct values of Scottish voters: 'Socialism, the idea of common ownership and control of the abundant wealth and resources that exist in our society, has always been popular in Scotland, it is just that there is no mainstream political party that articulates those ideas' (*The Herald* 10 April 1999). The SSP campaigned for a Scottish socialist republic and projected itself as a socialist alternative to New Labour, attempting to appeal to disillusioned traditional Labour supporters. The 100-point manifesto included collective ownership of transport, oil, finance, electricity and gas; progressive taxation; abolition of student fees and the reintroduction of grants; TV licenses and heating vouchers for pensioners; rejection of the Private Finance Initiative; phasing out of private health care; and a nuclear free Scotland.

Like the Greens, the Socialists were galvanised by the prospect of a fair voting system in the Scottish elections. Sheridan commented on the electoral system: 'The proportional representation element may be confusing, but it gives us a real chance. It's all we want. There has to be change' (*The Scotsman* 10 April 1999). Early in the campaign the media regarded the SSP as 'the most promising fringe party' of the elections (*The Herald* 3 March 1999; *The Scotsman* 4 March 1999).   An SSP commissioned System Three poll was widely quoted by the party and media; the poll suggested that 13% of Scots would consider voting for the party.

In the event, the SSP did well to challenge the suggestion that its support was 'ghettoised' in the Glasgow area (Milligan 1999). Although the Glasgow regional vote was by far the SSP's best performance, the party increased its vote in other parts of the country and attracted an average of 4.97% across the 18 constituency seats.   And the Hamilton South by-election at the end of September 1999, when the SSP candidate finished third, behind Labour and the SNP, reinforced the new party's position as a credible electoral force.[14]

The Scottish Green Party

The Scottish Greens have a 20 year history. For much of that time the party appeared marginal and ineffective. In 1990, a Scottish Green had briefly served as a regional councillor, in Highland, but the party's best election performance prior to 1999 was in the 1989 European Elections when it attracted 7.2% of the vote.[15] Robin Harper, an Edinburgh teacher, had been a party member since 1985. He had stood in three European Elections and the 1998 North East European by-election, two British General Elections, and in the Perth and Kinross Westminster by-election in 1995. On each occasion, he paid his own deposit; £500 for UK elections, £1,000 for the Euros. He lost every single deposit except in 1989 when he achieved 10.47% of the vote.

The Scottish Green Party programme in 1999 consisted of a wide range of environmental and social issues: pollution controls, a basic income scheme, a Scottish community bank to fund community businesses, local currency schemes; a housing investment scheme; the abolition of student tuition fees and the reintroduction of grants for the less well off; an integrated transport policy; and a ban on GM crops. The party was keen to stress at every available opportunity that it was not a single-issue party.

The Greens took the decision to put forward 41 regional candidates but no constituency candidates mainly because of financial constraints but also because they recognised that the proportional list system provided them with their best opportunity for election. The Green campaign was inventive, dynamic and media friendly. A decision to concentrate grassroots campaigning on the city centre streets of Edinburgh was combined with simple and effective media stunts and press releases. The party adopted the campaign slogan 'Vote Green 2' and used an enormous luminous Green '2' as an election prop. The campaign itself revolved around a few simple themes. The objective was to inform voters that they could vote Green on the regional lists. At every opportunity the party emphasised that 'Scottish Green Party' would appear on every peach coloured ballot paper in Scotland. And they constantly referred to 'around 6%' as the golden threshold beyond which they would be elected. Robin Harper urged Labour supporters who were concerned about the environment to consider giving their second vote to the Greens. At the same time, he attacked Labour for not delivering its promises on the environment. The Greens also attacked the Government for a poor public education campaign. The SGP argued that the official election material under-emphasised the fact that the new system could help new and small parties make an impact. The party launched its own voter awareness campaign and information hotline - a 'Second Vote hotline'.

The Greens entered the Scottish Parliament by the smallest of margins and the result was made all the more dramatic by problems at the count in Edinburgh. The count was suspended before its completion and the result was announced late in the afternoon of Friday 7th May. The returning officer Tom Aitchison, the city's chief executive, argued that the logistical task of counting 750,000 ballot papers was exacerbated by the length of the ballot paper for the second vote (*Press and Journal* 8 May 1999). The Edinburgh area is clearly the electoral base of Scottish Green support.[16] The party did not surpass 4% in any other region; the next best Green performance was in Glasgow, where the party attracted 3.96%. (The worst result for the Greens was in Central Scotland, with only 1.79%.) However, the Green vote did increase quite significantly across Scotland, and particularly in the Glasgow area. The party claims to be the fifth party in Scotland because it attracted more regional votes and more European Election votes than any other small party in 1999.

Other Parties

The Highlands and Islands Alliance, a small party of community and environmental activists, was formed to fight the Scottish elections and was widely predicted to do well. The Alliance attracted a disappointing 1.29% on May 6th, but the party was interesting for a number of other reasons. It set a precedent as the first party with 'job-sharing' candidates, with two people standing for one MSP place. While this was objected to by the Scottish Office and Henry McLeish, the Scottish Devolution Minister, the Alliance argued that job sharing was consistent with the objective of a 'family-friendly' Parliament. It quoted the report of the Scottish Parliament's consultative steering group, chaired by McLeish, which claimed that the Parliament would be flexible, responsive and family-friendly. The prominent face of the Alliance was Lorraine Mann, an anti-nuclear activist. She argued that she would be unable to work full-time as an MSP because of family commitments and because of her location, north of Inverness, and she pointed to the problems that would be encountered by many MSPs from rural areas.

Also in the Highlands and Islands area, Sir Iain Noble, the merchant banker, stood as an independent candidate. His candidacy was the closest thing to a 'Business Party' to emerge. He clearly indicated that he would not have stood under the old system: 'For several years I've been steadfastly non-political from a party point of view, but strongly in favour of a Scottish Parliament. The chance has now arisen because of the regional list system' (*The Scotsman* 5 March 1999).

It is clear that the creation of a Scottish Parliament spurred the small parties into action. The PR element provided hope for the non-major party candidates, and media exposure ensured an increase in their credibility. The success of the SSP and Greens confirmed this upward trend of credibility. However, the success of these parties was not predicted by the pundits. In the final week of the campaign, Malcolm Dickson (*The Herald* 23 April 1999) argued that the System Three polls showed little sign of small party support:

> Overall, it is a picture of continuing dominance of Scottish politics by the four major parties. Nowhere do any of the small parties look likely to make a breakthrough. This will come as a disappointment to the Scottish Green Party and the Scottish Socialist Party, in particular, who had been predicting gains for their cause. The Highlands and Islands Alliance, on the other hand, may have an outside chance.

The Relevance of the Small Parties in the Scottish Parliament

The limited proportionality of the Additional Member electoral system used in Scottish Parliamentary elections, with only 56 additional members, may mean that the number of SSP and Green representatives will not increase, or even be maintained. It will still be very difficult for these parties in forthcoming elections. Nevertheless, the new system does allow for multiple party loyalties on the part of the voter, and increasing voter sophistication is likely as the electorate comes to terms with the new system. Moreover, the initial electoral success of these parties is likely to have a positive effect on voter perceptions. Breaking through the electoral threshold has given these parties credibility and legitimacy. The significance of the election of Sheridan and Harper should not be underestimated. Their election also had an enormous impact on the parties themselves, including increases in membership, activism, and ability to conduct research. The Greens, for example, for much of their 20 year history have struggled with a number of part-time activists but now have a full-time research team.

Perhaps more important is the potential influence of these parties on policy making. It is possible that the small parties in the Parliament, along with Canavan, may be rather insignificant. They may not be 'relevant' in that they are unlikely to have coalition or blackmail potential (Sartori 1976). However, the existence of these parties in the Parliament is important for a number of reasons. There is no doubt that both the Socialists and the Greens are radical parties with alternative ideas and policies. The ideas they generate can only serve to represent a wider range of interests than the traditional four parties in Scotland. The presence of these parties helps to keep issues on the agenda.

Furthermore, because Sheridan and Harper (and Canavan) are independent voices they are free to ask pertinent and searching questions of the executive. Sheridan and Canavan in particular have proved to have considerable 'nuisance value' and this explains why the media are so interested in them. Indeed Sheridan, Canavan and Harper are much more likely to be quoted by the media than the major party back-benchers. Events which have sparked media interest include Sheridan's reluctance to swear allegiance to the Queen; attempts by the other parties (Liberal Democrats) to remove the three from their prominent seats in the Chamber; and a debate over whether the three would be allowed a seat on the committees.[17]

MSP status also brings with it certain legislative entitlements. As well as the right to question the executive in debates, each MSP is entitled to introduce two private members bills during the first four year parliament, which stand a much better chance of becoming law than their Westminster equivalents.[18] Sheridan introduced a bill to ban warrant sales for non-payment of debts, with the support of Labour and SNP MSPs. The individual MSPs can also influence legislation through their work on the committees.[19]

As well as enjoying powers as individual MSPs, Sheridan, Harper and Canavan have potential influence working with others. When voting on legislation they can form alliances with each other and with sympathetic rebels within other parties. Sheridan and Harper have already formed an effective alliance with each other and with Canavan. Indeed, Robin Harper and Tommy Sheridan both rushed to congratulate each other when elected and were united in their criticism of 'media indifference' and 'pundit complacency'.[20] These alliances mean that the main parties can't afford to ignore the smaller parties as they could conceivably hold a vital vote or two. It is certainly much easier now for the SSP and Greens to get other parties to engage in dialogue with them.

Conclusion

The 1999 Scottish Parliamentary elections have produced a more interesting party system in Scotland. The election has increased the diversity of parties, and the electorate can choose from a wider range of credible, legitimate parties. While the Scottish party system is far from highly fragmented, the new electoral system means that all parties are minorities in the Parliament: Mitchell (1999: 33) refers to the 'shifting coalitions of minorities'. Consequently, once a party enters the Parliament it gains formal influence, becoming a relevant part of the legislature. The SSP and the Greens have gained a foothold in Scottish politics. As individual MSPs, and as members of broad opposition alliances, they can make a difference.

Notes

1   For example, Drucker (1982: 30) notes that the Social Democratic Party (SDP) was never as popular in Scotland.

2   Defining the term 'small party' is an arbitrary process (See Müller-Rommell and Pridham 1989). Various different terms are used indiscriminately in the literature, including minor parties, fringe parties, peripheral parties, flash parties and pressure parties. In this instance small parties are treated as those that exist on the periphery of Scottish politics, never having gained electoral representation at Westminster, although they may have had some experience of electoral representation at the local level. In practice, small parties are treated as all parties other than the four major parties.

3   The impact of the new constitutional arrangements on the small parties themselves - their election experiences, organisational implications (membership/ activism) and so on - is the subject of ongoing research by the author.

4   Deposits in General Elections are prohibitively high for many small parties and independent candidates. The Representation of the People Act 1918 determined that all parliamentary candidates had to deposit £150 with the returning officer. This money would be returned if the candidate received one eighth of the votes cast. In 1985 the rules were changed: deposits were raised from £150 to £500, because so many fringe candidates were standing, but the threshold at which the deposit was returned was lowered from one eighth to one twentieth (5%).

5   The SSA attracted an average of 1.83%.

6   The cost of standing in the constituencies was the same as a UK Westminster election (£500). The rules governing the return of deposits, in both the constituencies and in the regions, were the same as in UK General Election (5%). As before, all candidates were entitled to state funds for mailing of election

leaflets. The candidate/party produces leaflets but the state covers the postage costs of leaflets sent to every member of the constituency/region.

[7] This amounted to more than 300 small party candidates in 1999.

[8] Mid Scotland and Fife had only nine lists, the smallest number of all the regions. Glasgow had 14 lists.

[9] Scotland was treated as a single region, the cost of an election deposit was £5,000, and the deposit threshold was 2.5%. In previous European Elections, the deposit had been £1,000 per seat, with 8 Scottish seats in total.

[10] Two Green MEPs were elected in England - Jean Lambert in London, and Caroline Lucas in the South East region - along with two UK Independence Party candidates in the Eastern and South West areas.

[11] European Elections have a reputation for encouraging high levels of protest voting. Voters appear willing to turn their backs on the big four because these elections are often viewed as rather unimportant, or 'second order' (Reif 1985). The relative success of the Greens in 1989 is often cited as an example of these election effects.

[12] The SSA put forward 16 candidates in 1997 and attracted 9,508 votes, 1.8% of votes cast.

[13] Tommy Sheridan describes himself as party convener, not leader. Hugh Kerr is a member of the SSP but is an MEP for the *English* seat of Essex West and Hertfordshire East.

[14] The Liberal Democrat candidate lost her deposit and finished sixth, behind supporters of Hamilton Academical Football Club.

[15] The UK Greens however achieved nearly 15% of the vote.

[16] There is some evidence to suggest that the East of Scotland is more environmentally aware than the West (McCormick and McDowell 1999: 55).

[17] It was claimed that the three were not entitled to a seat on a committee under parliamentary rules, as the committees are supposed to represent the balance of power between the parties. However, the business bureau created spaces for them. Harper, for example, specifically requested a seat on the Transport and Environment Committee.

[18] The bills require nomination by one member and the signatures of 11 other members.

[19] Indeed the committees have the power to introduce legislation.

[20] Interviewed in the *Sunday Herald* (25 April 1999) Robin Harper argued that if elected he would seek to forge alliances with members of other parties on specific issues: 'I have met so many people from different parties, on platforms and so on, and I find them agreeing with me. They say, as Robin says, poverty is a major cause of ill health, or, as Robin says, public transport needs massive investment and we have common ground.'

References

Butler, David and Butler, Gareth (1994), *British Political Facts*, MacMillan, London.

Denver, D. and Bochel, H. (1994) 'The Last Act: The Regional Elections of 1994', *Scottish Affairs*, no 9, Autumn 1994, pp.68-79.

Drucker, Henry (1982), 'The Curious Incident: Scottish Party Competition Since 1979', in David McCrone (ed.), *The Scottish Government Yearbook 1983* Research Centre for Social Sciences), Edinburgh, pp.16-32.

Milligan, Tony (1999), 'Left at the Polls: The Changing Far-Left Vote', *Scottish Affairs*, No.29, Autumn 1999, pp.139-156.

Mitchell, James (1999), 'Consensus: Whose Consensus?' in Gerry Hassan and Chris Warhurst (eds), *A Different Scotland: A Modernisers Guide to Scotland,* The Big Issue/ Centre for Scottish Public Policy, Glasgow, pp.28-33.

Müller-Rommel, Ferdinand and Pridham, Geoffrey. (1989), *Small Parties in Western Europe: Comparative and National Perspectives*, Sage, London.

Reif, Karl-Heinz (ed.) (1985), *Ten European Elections,* Gower, Aldershot.

Rüdig, W., Franklin, M.N. and Bennie, L.G. (1996), 'Up and down with the Greens: Ecology and Politics in Britain, 1989-1992', *Electoral Studies*, vol.15, no.1, pp. 1-20.

Sartori, Giovanni (1976) *Parties and Party Systems*, Cambridge University Press, Cambridge.

PART III
INTER-GOVERNMENTAL
RELATIONS: SCOTLAND-UK

7 'Layers of Democracy': Making Home Rule Work

JOHN FAIRLEY

Introduction [1]

The election of the first democratic Scottish Parliament on 6th May 1999, changed Scottish and UK politics. Indeed the desire for change - in particular the demand for more democracy within Scotland, and the UK Government's determination to decentralise power - were powerful 'change drivers'. The introduction of a national parliament, with the promise of further devolution within Scotland, created a need for mechanisms to manage the workings of the new democracy.

Multi-Layer Democracy

This was the phrase employed by the Scottish Affairs Select Committee of the UK Parliament, in its thoughtful report (SAC 1998) on the new democracy. The Committee was concerned to examine how the different layers might best work together and it also reflected on the political values which would need to be developed and observed if the system was to work well. The main layers of this new democracy are:

A.   8   Members of the European Parliament (MEP)

B.   72   Members of the UK Parliament (MP)

C.   129   Members of the Scottish Parliament (MSP)

D.   1,222   Members of 32 Local Governments

This chapter will focus first on layers B and C, the Scottish and UK Parliaments. It will then examine, in a little more detail, the relations between the Scottish Parliament (C) and local government (D), the two layers which are internal to Scotland and which, between them, provide over 94% of the elected politicians who comprise the new democracy.

Home Rule or Devolution?

The constitutional change that is underway in Scotland, and in Scotland's changing relationship with the UK, is unique in that it is a response to the pressures of nationalism, and in particular to the Scottish National Party. The change process is capable of being understood in competing ways. The manner in which the change is understood is likely strongly to influence the approach which is taken to the management of the new democracy.

MacCormick (1998) has clearly set out the dominant view of the UK constitution, together with his own, different view of the 'Scottish anomaly'. The former view is likely to predominate within UK institutions, and in the thinking of the politicians of the Unionist parties, particularly but not exclusively at Westminster.

The dominant view of the UK constitution, prior to the implementation of the Scotland Act of 1998, is that in 1707 Scotland was brought within the scope of the evolving English constitution. Within this 'incorporating' Union, a unitary state was created under the supervision of Westminster. Scottish distinctiveness was recognised by the Westminster Parliament, the legislative basis of the Union allowing for Scottish autonomy in matters of religion, education, law and local government (MacCormick 1998). Within the then very limited sphere of government, Scotland largely looked after itself, conducting its domestic affairs with considerable autonomy from the rest of Britain (see Paterson 1994), with local authorities being particularly important.

How does this dominant view of the UK regard the current change process in Scotland? The answer is likely to be that devolution is the preferred policy of the Westminster Government on which it consulted with the people of Scotland through the device of the two-question referendum in September 1997. Once the Government's intentions had been legitimised by the Scottish people, legislation was brought forward to create the devolved Scottish Parliament. However, the initiative, and the crucial impetus, came from Westminster, and the Westminster Parliament remains sovereign, as the 1997 White Paper and the 1998 Scotland Act explicitly confirmed (Bogdanor 1998: 288; MacCormick 1999: 59). In the social science devolutionary jargon, the 1998 Act confirmed Westminster as 'superintendent', and Edinburgh as 'subordinate' (Sabel 1997).

In creating devolved government for Scotland, Westminster primarily offered Scotland a means of providing democratic scrutiny over its existing areas of policy autonomy. As MacCormick (1998: 143) puts it: 'In place of managed quasi-federalism, there will be democratic quasi-federalism.' However, it is also important to note that the Scotland Act extends the sphere of Scottish autonomy, making the Parliament considerably more powerful than the German lander (SAC 1998, Vol. 2: 42), and that the 'retaining model' (McFadden and Lazarowicz 1999) provided by the Act offers scope for further developments in the powers and competencies of the Parliament.

The dominant view of constitutional change seems likely to lead to a particular view of how multi-layer democracy is to be managed. In particular it would be expected to seek approaches to political management which assert and reinforce a hierarchy of politicians. Indeed, it could be argued that simply setting out the 4 layers, A to D, as above, implicitly accepts such a hierarchy. Within this view, there would be some tension between those who think that the EU is now dominant, and the so-called 'eurosceptics' in the Labour and Conservative Parties who would continue to assert that it is Westminster MPs who are at the top of the hierarchy.

At the time of writing (December 1999) this hierarchical view has emerged very sharply, and in ways which are potentially damaging to the young Scottish Parliament. Labour Westminster MPs have asserted their right to express a view in a number of policy areas which are devolved, and have questioned the legitimacy of list MSPs[2] who seek to represent constituents. In this process the primacy of Westminster is made clear, and it is further asserted that there is hierarchy within the Scottish Parliament, between those who are legitimised, and therefore made superior, by direct election, and, on the other hand, those who must be seen as second class because they are merely elected by the proportional route. The Presiding Officer of the Parliament was forced to move quickly and publicly reassert the equality between MSPs of both types.

An alternative view of the constitutional change process is no less significant in its implications. The modern UK may be understood as a 'union state' (Bogdanor 1997: 30) rather than a unitary state. The union was built by a variety of processes, including in the case of Scotland, treaty. The coming together of Scotland and England was 'a political bargain' (ibid, 25), between two independently and differently sovereign states[3] (MacCormick 1998), in a relationship which always retained the potential for renegotiation (Nairn 1998a).

The Convention Scheme was perhaps the most important 'moment' in the process because it changed devolution from an abstract principle into something concrete and practical on which the electorate could vote (MacCormick 1999: 199). The 1998 Scotland Act represented the legislative formalities of this renegotiation. The process critically involved:

the development within Scotland of the widely-supported 'Convention Scheme' (SCC 1995) for the design of the Parliament; the election of the pro-devolution New Labour government in the May 1997 UK elections; the referendum on the Convention proposals in the form of a White Paper in September 1997; the passing of the Scotland Act; and, the first Scottish General Election in May 1999. The 1998 Act is thus understood not as 'devolution from on high', but rather the Westminster Parliament legislating to provide the new settlement which Scotland designed and demanded, a settlement which could further tip the UK's asymmetry in Scotland's favour (Bogdanor 1997: 33).

This view, of 'the partnership renegotiated', would imply the need for new and more positive relationships between elected politicians, based not on old concepts of hierarchy, but rather on the recognition of a 'democratic division of labour' and the need for greater 'parity of esteem' between those elected to the different layers in the system. The likelihood is that this more positive view of the required political relationships will at least face competition from the historically dominant paradigm. However, the adherents of the more positive view now have the framework of the new settlement within which to work, and a new context where 'the requirement for co-operation will generate its own effective working practices in the short term' (Jeffrey 1998).

The Scottish Affairs Committee (SAC 1998) considered the views of expert witnesses on Germany and Spain. It paid particular attention to the German principle of 'federal comity or loyalty', which underpins Federal-Lander, and Lander-Lander relations. The principle, as defined by the Constitutional Court, is that

> A federal state can exist only if the federation and the Lander in their relations to each other take into account that the standards under which they make use of formally existing competencies are governed by mutual consideration (Leonardy 1998: 37).

The UK is not a federal state, in the sense that it lacks the institutional framework by which federalism is usually defined. However it is possible to see the Scottish Parliament and the Welsh and Northern Ireland Assemblies as developments in that direction. And there are currently no proposals in the UK for the complex of decision-making relationships, which underpin German federalism (ibid). Indeed the idea of 'whole state' consensual, decision-making, within which the federation and the lander are equally represented, would probably be viewed as incompatible with, or undermining of, the principle of Westminster sovereignty.

It is also possible to view the current changes from a different theoretical perspective which seeks to ground federalism in society, and in social practices, rather than in institutions (Paterson 1994: 20). On this view federalism is created and sustained by the collaborative practices of economic, political, social and cultural life. The Parliament can then be seen as an institutional and democratic form, which is much more relevant to, and reflective of, modern Scotland.

The actual mechanisms which evolve to secure the management of the new democracy are likely to do so under two sets of pressures: first, what is required to make the new system of government work; and, second what other devices are possible and politically acceptable in the new context of extended democracy. Perhaps we may expect the first category to be more consensual, and the second to be more characterised by discussion and disagreement.

Mechanisms for Achieving Co-operation and Mutual Consideration

The Political Parties

The Labour Party, as the party of government in Westminster and the largest party in Scotland's coalition government, seems likely to seek policy consistency in England and Scotland, at least in the UK government's priority areas where there is devolved responsibility, education, for example. Clearly this would be more difficult to achieve in a situation where different parties controlled the parliaments.

Perhaps the most important set of mechanisms is provided by the political parties and their modes of conduct, towards each other and internally. Following the clear outcome of the 1997 Referendum, and its acceptance by the Conservative Party, all of the significant political forces in Scotland supported the Parliament. All of the parties present in the Edinburgh Parliament, with the exception of the Greens, also have a presence elsewhere in the multi-layer system. The main parties also have MPs in the Westminster Parliament, although the Conservatives do not have any members elected from Scotland.

The political parties then, in their internal communications, their candidate selection procedures, internal briefings, policy-making processes, conferences etc can take the lead in promoting the virtues and the values of Home Rule, and of the consensual approach to politics which was promoted by the Convention Scheme, and which, to some extent, was built into the Parliament's operating principles. Clearly this shift, in some important aspects at least, has occurred within Scotland. The most difficult aspect, given the UK's adversarial traditions, is for the parties to learn and assimilate the new values of proportional and more consensual politics. However, at the time of writing, there are very encouraging signs in the conduct of the Parliament and the work of its committees.

Similar developments also need to occur outside of Scotland in the British-wide political parties, as they seek to catch up with the new institutional and democratic diversity of the UK. And there may be real difficulties here. The British parties are not represented in the Northern Ireland Assembly. The consensus supporting the Welsh Parliament was weak compared with the case of Scotland. The British political parties may have a dilemma. Having traditionally presented themselves as rather uniformly 'British', they now need quickly to present themselves as well-informed and positive regarding the new Parliament and Assemblies. Failure in this regard will raise the likelihood that Westminster politicians are perceived to be out of touch with, or even irrelevant, to Scotland in particular. Within the first 6 months of the Parliament there were cases of MPs and the heads of cross border public authorities being ridiculed in the Scottish media for their lack of knowledge of current developments. However, if the British parties embrace national and regional diversity too enthusiastically, then they run the risks of further fuelling devolutionary pressures, and perhaps more importantly, endorsing the internal expression of these differences in ways which would make party management and policy consistency more difficult to achieve. Indeed, if they were to become too enthusiastic, then it is possible that their 'Britishness' could be called into question.

And the nationalist parties also have their difficulties to address. They need to develop some new grounds on which to reassert their claims to leadership in their respective countries, and, at the same time maintain some positive presence in Westminster and in Europe.

Concordats

In order for the different layers of democracy and institutions of state to function together or at least in ways which are consistent and compatible, there need to be formal agreements on roles and procedures. The principal Concordat, was issued from the office of the Lord Chancellor, where formal UK ministerial responsibility resides, in October 1999 (Lord Chancellor 1999). The Concordats are agreements on 'the principles which will underlie relations' between 'the UK Government and the devolved administrations in Scotland and Wales.' They are not legally binding agreements.

The principal is the Memorandum of Understanding (MoU) which established the device of the Joint Ministerial Committee (JMC) of the Scottish and UK Parliaments, which met for the first time in December. Alongside the JMC four 'overarching Concordats' were set out to cover 'the handling of matters with an EU dimension; financial assistance to industry; international relations touching on the responsibilities of the devolved administrations; and statistical work across the UK.'

As agreements which are not legally binding, the Concordats have, to date, been little discussed. They are unlikely to prove controversial unless they come to be seen as obstructing the Scottish Parliament in some way. Their manner of development, which might have been expected to be controversial, has passed largely without critical comment.

There was a general expectation that the Concordats would be based on discussion and agreement between the Scottish and UK administrations. The then Scottish Office (1998), in a Memorandum, appeared to be unequivocal when informing the Scottish Affairs Committee that, 'Concordats will be published in draft when they are ready.'

However, the principal Concordat simply appeared, from Westminster, apparently without any input from the coalition MSPs. Later in 1999 the process of developing Concordats to cover the related activities of Whitehall and Edinburgh Departments began, the first of these being announced, by the Lord Chancellor, during a visit to Scotland.

At the time of writing it is impossible to say how the Concordats will be used and whether they will be effective. The JMC offers a forum for the discussion of Scottish concerns, which was not previously available. The agreements themselves appear, for the most part, to avoid being precise or prescriptive, which suggests that they may be capable of providing frameworks for discussion, and where necessary, negotiation.

Disputes

As the SAC (1998 Vol. 1: xvi) pointed out

> There is considerable and legitimate potential scope for disagreement between state and sub-state (and even between sub-states) even where both administrations are of the same political complexion, but any disagreements will be exacerbated when they are not.

The 1998 Scotland Act provides mechanisms to resolve possible disputes over *vires* (ie whether or not the Parliament is acting within its powers). At the beginning of the legislative process, a Bill requires a written statement that it falls within the Parliament's competence. The Presiding Officer has to ensure that, during its passage, the Bill remains within the Parliament's available powers. Following its successful passage through the Scottish Parliament, a Bill faces a delay of four weeks before receiving Royal Assent. This delay is to allow the UK Government to check on the Bill's legitimacy.

The Scottish Law Officers or the Attorney General may refer a Bill to the Judicial Committee of the Privy Council. The UK has no Constitutional Court, and the Judicial Committee acts as a Court of Appeal for some countries of the Commonwealth. The Committee which will consider *vires* in relation to legislation passed by the Scottish Parliament will consist of five judges, at least two of whom will be Scottish.

Some experts doubt that this Committee will prove to be effective. Jeffrey (1998) argued that it would not have the authority of, for example, the German Constitutional Court. Nairn (1998b) thought that it would be unable to resolve political disputes between the Scottish and UK parliaments, and the SAC itself (1998: xvii) pinpointed tourism and inward investment as areas which could generate conflicts of a kind which the Committee would not be able to resolve.

Bogdanor (1999) has pointed out that the 1920 Government of Ireland Act contained a much stronger declaration of Westminster sovereignty over the UK's first devolved administration than is to be found in the Scotland Act of 1998. However, in practice the UK Government found that it could not enforce its supremacy. When disputes arose the UK Government gave way to the Government of Northern Ireland.

> If Westminster found itself incapable of exercising its supremacy over the Northern Ireland Parliament, how much more difficult its task will be *vis-à-vis* Scotland. For Northern Ireland did not see herself as a separate nation within the United Kingdom, nor had she sought devolution. (Bogdanor, 1999: 289)

The Forgotten Layer: Local Democracy

At the time of writing it remains unclear precisely how the new Parliament will affect local government. The Scottish Parliament now oversees almost every aspect of local government.

In his submission to the SAC (1998, Vol. 2: 27) Jeffrey argued that

> The Scottish Parliament's responsibility for local government will be an important indicator of the qualities of post-devolution governance in Scotland. When formerly centralist states decentralise power to the regional level, local government can suffer from the 'decentralisation of centralism' (as has, for example, been the case in Belgium). And, as the

case of Germany shows, local government autonomy can be 'squeezed' even in states with a long decentralised tradition.

Heywood (1995: 157) points to a similar process in Spain, and explains it in terms of the new regional governments not wanting to risk losing any power which could reduce their status *vis-à-vis* central government. Bogdanor (SAC 1998:xx) pointed out that the Government of Catalonia attempted to abolish local government entirely.

In Scotland, the debate on the likely implications of a Scottish Parliament really began to get off the ground around the publication of the Convention Scheme in 1995. Midwinter (1995) thought that a Parliament would be likely to weaken local government by taking powers to the centre. Rosemary McKenna, the then President of the Convention of Scottish Local Authorities (CoSLA), addressed the issues in a Conference speech early in 1996. Douglas Sinclair (1997), the then Chief Executive of CoSLA, discussed the dangers for local government, and took the view that devolution was likely to be positive.

In addition to considering local government as a whole, a number of commentators have looked at the implications of the Parliament for particular local government services. Examinations of education (Fairley 1998), economic development (Fairley 1999) and planning (Hayton 1999) are examples from a wide-ranging discussion.

Local Authorities and the Convention

Local authorities supported the Convention by joining, and by providing some resources. Most of the local authorities, 58 out of 65, were members, and without the material support provided, it must be doubtful whether the Convention could have completed its scheme in time to win the full backing of the British Labour Party in its 1997 election campaign. Without this local government support, the implementation of devolution may have been slower, and less consensual (Fairley 2000a).

The Convention (SCC 1995: 17) advocated 'subsidiarity' within Scotland, and envisaged the Parliament safeguarding and, where possible, extending the roles of local government. The White Paper (Scottish Office 1997: 19) made it clear that the UK Government was opposed to the Parliament overseeing centralisation within Scotland:

> In establishing a Scottish Parliament to extend democratic accountability, the Government do not expect the Scottish Parliament and its Executive to accumulate a range of new functions at the centre which would be more appropriately end efficiently delivered by other bodies.

The Scotland Act does not deal with local government, thereby giving the Parliament a free hand. The Parliament could seek to decentralise and devolve power within Scotland, and to give new roles to local government, in the spirit of the Convention Scheme and the White Paper. Equally, however, it could be a centralising force within Scotland, taking powers away from local authorities and reducing their importance.

The Politics of Local Government

The UK's local government systems are weak, compared with many EU countries, in that they are tightly controlled from the centre, and have little autonomy in relation to policy or finance. Within the UK, Northern Ireland has the weakest system. One test of the Scottish Parliament will be whether it acts to strengthen local government, and the general policy stance will be one factor which helps to shape relationships between the two layers.

Table 7.1 Outcome of the 1999 Local Government Election

Party	No Seats	% seats	No Councils
Scottish Labour	551	45.1	20
Scottish National Party	204	16.7	2
Independent/other	204	16.7	7
Scottish Liberal Democrats	155	12.7	0
Scottish Conservatives	108	8.8	0

Source: SLGIU 1999
Notes: Three councils have no overall control. Five Labour and one SNP councils are minority administrations. Four councils are Independent, and three are coalitions of Independents with Labour (2), or Liberal Democrats (1)

While Scottish local government is weak in certain respects, and while it has recently lost important functional responsibilities such as water and sewerage services, nevertheless in certain respects individual local authorities are arguably stronger than before. The 1994 Local Government etc. (Scotland) Act 1994, abolished the two-tier system of regional and district councils on mainland Scotland, and, in its place put a network of 29 'unitary' authorities. These councils have the advantage that each is the only elected body for its territory. In the new system there can be no doubt over which institution is the legitimate local democratic voice, and this should help to simplify relations with the Parliament.

As in the Edinburgh-Westminster relationship, politics and the political parties will be of central importance. The 1999 elections saw the largest number of candidates, 3,928, contest the smallest-ever number of local government seats, 1,222. Professor Stewart (1998) has argued that 'there are probably fewer councillors in Scotland than in any other country in Europe', pointing out that in Scotland there is one councillor for over 4,000 people, while in Europe the average is a ratio of 1:400. In the 1999 elections, Labour lost some 50 seats, but emerged by far the most important party, with 45% of the seats, and controlling 20 of the 32 councils.

Mutual Consideration?

The fact that Labour is so dominant in Scotland's multi-layer democracy should, on the face of it, help the new arrangements to settle down, whether the focus is on Edinburgh-London, or local government and Edinburgh. Nevertheless, there remains considerable scope for disagreement and even conflict. At Westminster and in Edinburgh, Labour is very much 'New Labour', committed to a 'modernising' agenda for the public sector. However, in many of Labour's urban councils, 'Old Labour' is in control, and it is generally suspicious of or hostile to the 'modernising' agenda. In addition, in Scotland Labour has to govern in coalition, as is the case at the time of writing, or as a minority administration. In short, it cannot simply impose its own policies, rather it will have to negotiate, particularly perhaps, with those political forces that are stronger in rural Scotland, notably, the Liberal Democrats, the SNP, and the Independents. The Labour Party is weak in rural local government (Fairley 2000b), and it remains a challenge for the Parliament is to have positive relationships with all 32 councils - urban, rural and island. A further source of difficulty is that the main British political parties have traditions of mistrust or hostility towards Scottish local government. The Conservatives learned this attitude when they were confronted with local authorities that were unwilling to implement their policies. In the case of Labour centralist attitudes are perhaps more fundamental. The Liberal Democrats are, by tradition, more positive in their attitude to local government, but it remains unclear whether they will be able to influence the governing coalition in this policy area.

However, there have been two developments, which may help to foster a greater degree of parity of esteem between MSPs and local politicians. The first is that the elections to the parliament and to local government attracted very similar participation rates. This is important because it means that no layer of politicians is able to lay claim to a superior democratic mandate. The second point is that 31 of the 129 MSPs were previously local councillors, and a further 12 have local government experience. Three of the coalition Ministers were previously council leaders. This means that, at least in the first Parliament, there is likely to be well-informed debate about local government. However, it could be argued that, in the absence of a 'dual mandate' which would allow politicians to hold local and national office at the same time, the level of knowledge held by MSPs is likely to become dated and to decline in importance.

The McIntosh Report

The UK Government established an independent Commission, convened by Neil McIntosh, to look into relations between the Parliament and local government, and to report to the first Minister. The Commission reported in 1999, and, at the time of writing, the Scottish Executive is consulting again on its initial response to the report (McIntosh 1999).

McIntosh proposed a complex set of inter-related recommendations, political, organisational, managerial and financial, although local government finance lay outwith the Commission's remit. At the time of writing it is already clear that the Executive will not implement McIntosh in full. It has already rejected out of hand the proposal for an urgent review of local government finance, although it is possible that the Local Government Committee of the Parliament may take up this matter.

The Scottish Executive has so far shown itself to be most enthusiastic about two aspects of local government reform: first, electoral reform. A Commission has been set up under the chairmanship of Richard Kerley, to report on the most appropriate form of proportional representation for local elections. This is not simply a technical matter. Any new system is likely to be unpopular with urban Labour, which benefits most from the disproportionality of the present system. And any new system will have to be able to accommodate the complexity and diversity of local politics, including the importance of the Independents in island and rural Scotland. Second, the Scottish Executive is keen to continue the 'modernisation' of managerial decision-making in local government, and, in the tradition of UK governments, its thinking is primarily influenced by its perception of local government practice in the USA. The Executive wants local authorities to streamline their decision-making and to improve their managerial leadership. It also believes that Scotland's cities should have directly elected Provosts[4]. While there is a lot of support within local authorities for the first two types of change, McIntosh found almost no support for the direct election of Provosts.

A Local Government Concordat?

The McIntosh Commission raised the question of a formal agreement, or Covenant, being established between the Parliament and the local authorities, a proposal, which enjoys the support of the Convention of Scottish Local Authorities. The Executive has responded positively on the principle and is currently consulting on the best way to implement it.

In order to be effective, this agreement would need to address a number of issues. In addition to clarifying the local roles and responsibilities of councils, it would need to provide greater stability than has previously been available. This could perhaps be achieved by having a long-term agreement on the principles underpinning the relationship. At the very least an agreement would need to cover the 4-year term of the Parliament. It would need to set out local government's role in the governance of Scotland, and mechanisms for local government to work with the Parliament in the appropriate policy areas. And it would need to affirm and seek to take forward the principles of the European Charter of Local Self Government, which was recently adopted by the UK Government. As the Convention argued (SCC 1995: 17), 'This will allow local government the flexibility it wants and needs to act in the interests of its citizens in its provision of high and improving standards of service.'

Privatisation - a Constitutional Issue?

This aspect of change lies beyond the remit of the McIntosh Commission, and may simply be an aspect of 'policy continuity', namely the extension of the privatisation programme which was initiated by the Conservatives during the 1980s. This approach to policy favoured decentralisation through quasi-market measures rather than through democratic structures. The constitutional importance of this is that it will inevitably reduce the role and scope of local government and local democratic accountability, and, in so doing, it may go against the spirit of the Convention scheme. At the time of writing, the Executive wishes to transfer council-owned housing to community-based trusts. While most councils probably would not choose this option, it is made very attractive to them, because it is the only device available for bringing in private investment capital and because it will allow the transfer of housing debt. The new trusts will involve tenants in some aspects of management, ensuring a limited degree of accountability to service users.

However, the broader, traditional forms of democratic accountability will be lost. In education, the largest local authority service, current proposals would bring private finance in to the building of new schools and the upgrading of existing buildings. One consequence of this is that the private investors will then manage the buildings and employ the non-teaching staff. The proposals do not seem to contain any mechanisms for ensuring accountability to service users, namely students, their parents and the teachers. However, they will, if implemented, remove educational assets and staff beyond the possibility of local democratic influence.

There are two major ironies in these aspects of privatisation. First, very similar proposals by the Conservatives, particularly for education, were bitterly opposed in Scotland, and by local authorities. This opposition went hand-in-hand with the campaign to establish a Parliament that would be more responsive to Scotland's policy preferences. Current proposals would see the implementation of privatisation on a scale and in forms which the Conservatives probably would not have thought possible to achieve, at least in the short run. While the Scottish electorate appears clearly to have voted for the Parliament because it was expected to improve the Welfare State (Surridge et al 1998), the Executive seems intent on linking modernisation to privatisation, at least where the capital programme is concerned. The second irony is that these new policies strike at the heart of local government's traditional roles in housing and in education, and these are the two local welfare services in which the public takes most interest. At a time when there is much discussion about making local government more interesting, encouraging participation, and improving democracy, the proposed reforms will reduce the role of local government in key community services, and will reduce the scope of local democracy.

Barriers to Further Devolution?

There is no doubt that the Convention scheme envisaged a devolved and decentralised Scottish democracy. However, it is important to note that there are important cultural and political obstacles that could obstruct these processes.

When the Convention published its scheme for devolution, Scotland was perhaps the most centralised polity in Europe. The Scottish office controlled local government in most respects, and at times in its detailed workings. Similarly, the 'non-elected state' made up of the Non Departmental Public Bodies (NDPBs) and other local spending bodies, was tightly controlled from the centre.

Within this polity the central state simply did not trust local authorities. There was no question of regarding them as 'partners' in government. Local government had generally come to accept this state of affairs, and to look to, and depend upon the Scottish Office, and its NDPBs, to take the policy initiatives to which it would respond. There are two main issues which arise from this over-centralisation.

First, any attempt to implement the Convention's radical vision would require a major change in the values, attitudes and practices of the civil service and of local government. Second, it may be that local government simply lacks the 'policy capacity' which significant democratic devolution requires. It is common for even quite large local authorities to have 'policy units' of only one or two professional staff, and most of their time is occupied responding to Scottish Executive consultations, rather than on developing locally-grounded policy thinking.

Conclusions

The mechanisms for ensuring effective democratic partnership between the Scottish and UK governments began to be developed within months of the Scottish general election. These mechanisms will continue to evolve, as the Scottish Parliament becomes an effective legislature, and as the UK Parliament comes to terms with the consequences of devolution. It seems likely that these mechanisms will lead to further powers and competencies being devolved to the Scottish Parliament.

The implementation of the devolved democracy envisaged by the Convention looks much more difficult to achieve. The enthusiastic implementation of the McIntosh proposals would herald a start to this process, but it seems clear that this is not what the Scottish Executive intends. And, even with McIntosh in place, there would remain the great challenge of re-thinking Scotland's tradition of very centralised governance.

Notes

[1]   I received helpful comments on an early draft of this paper from Professor Lindsay Paterson.

[2]   73 MSPs were elected on the traditional first-past-the-post system. They tend to be known as 'constituency MSPs'. The remaining 56 were elected from closed regional lists which were designed to correct the disproportionalities arising from first-past-the-post. The second group are known as 'list MSPs'.

[3]   The English tradition regards the Westminster Parliament, or 'the crown in Parliament' as sovereign, while in Scotland there is a doctrine of 'popular sovereignty' which dates from 1320 (See MacCormick 1998).

[4]   Scotland has a long tradition of Provosts being elected from the Council.

References

Bogdanor, V. (1997), *Power and the People - A Guide to Constitutional Reform*, Victor Gollancz, London.

Bogdanor, V. (1999), *Devolution in the United Kingdom*, Oxford University Press, Oxford.

Fairley, J. (1995), 'The Changing Politics of Local Government in Scotland', *Scottish Affairs*, Winter.

Fairley, J. (1998), 'Local Authority Education in a Democratic Scotland', *Scottish Educational Review*, May.

Fairley, J. (1999), 'Economic Development: the Scottish Parliament and Local Government', in J. McCarthy and D. Newlands (eds) *Governing Scotland: Problems and Prospects; the Economic Impact of the Scottish Parliament*, Ashgate, Aldershot.

Fairley, J. (2000a), 'Local Government', in K. Dixon (ed) *L'Autonomie Ecossais,e* ELLUG, Grenoble.

Fairley, J. (2000b) 'Scotland's New Democracy - New Opportunities for Rural Scotland?', *Scottish Affairs*.

Heywood, P. (1995), *The Government and Politics of Spain*, MacMillan, Basingstoke.

Jeffrey, C. (1998) 'Evidence to the Enquiry into the Operation of Multi Layer Democracy by the *Scottish Affairs Committee*', HC 460-ii, The Stationery Office, London.

Leonardy, U. (1998), 'Memorandum to the Enquiry into the Operation of Multi-Layer Democracy by the *Scottish Affairs Committee*', HC 460-ii, The Stationery Office, London.

Lord Chancellor (1999), *Devolution: Memorandum of Understanding and supplementary agreements between the United Kingdom Government, Scottish Ministers and the cabinet of the National Assembly of Wales*, Cm4444, The Stationery Office, London.

MacCormick, N. (1998), 'The English Constitution, the British State and the Scottish Anomaly', *Scottish Affairs*, Special Issue, Understanding Constitutional Change.

MacCormick, N. (1999), *Questioning Sovereignty*, Oxford University Press, Oxford.

McFadden, J. and Lazarowicz, M. (1998), *The Scottish Parliament, an Introduction*, T&T Clark, Edinburgh.

McIntosh, N. (1999), *Moving Forward; Local Government and the Scottish Parliament*. The Report of the Committee on Local Government and the Scottish Parliament, Edinburgh, the Scottish Office, June.

McKenna, R. 'A Scottish Parliament: Friend or Foe to Local Government?' in L. Paterson 1998 (ed), *A Diverse Assembly - The Debate on a Scottish Parliament*, Edinburgh University Press, Edinburgh.

Midwinter, A. (1995), *Local Government Reform in Scotland - Reform or Decline?*, MacMillan, London.

Nairn, T. (1998a), 'Virtual Liberation or: British Sovereignty Since the Election', *Scottish Affairs*, Special Issue, Understanding Constitutional Change.

Nairn, T. (1998b), *Evidence to the Scottish Affairs Select Committee*, The Operation of Multi-Layer Democracy, HC 460, the Stationery Office, London.

Paterson, L. (1994), *The Autonomy of Modern Scotland*, Edinburgh University Press, Edinburgh.

Sabel, C. (1997), 'Constitutional Orders: Trust Building and Response to Change', in J. R. Hollingsworth and R. Boyer (eds), *Contemporary Capitalism; the Embeddedness of Institutions*, Cambridge University Press, Cambridge.

SAC (1998), 'The Operation of Multi-Layer Democracy', HC460, *Report of the Scottish Affairs Committee*, the Stationery Office, London.

SCC (1995), *Scotland's Parliament: Scotland's Right*, Scottish Constitutional Convention Edinburgh.

Scottish Office (1997), *Scotland's Parliament*, The Scottish Office, Edinburgh.

Scottish Office (1998), 'Further supplementary memorandum from the Scottish Office' (13.10.98), published *in Scottish Affairs Committee 1998 The Operation of Multi-Layer Democracy, Volume 1*, HC 460-I, p.9, The Stationery Office, London.

SLGIU (1999), *Bulletin 113*, Glasgow: Scottish Local Government Information Unit.

Stewart, J. (1998), 'Strengthening Local Democracy in Scotland: the Challenge to Local Government', *Scottish Affairs*, no. 25.

Surridge, P., Paterson, L., Brown, A. and McCrone, D. (1998), 'The Scottish Electorate and the Scottish Parliament', *Scottish Affairs*, Special Issue, Understanding Constitutional Change.

8 Quasi Government in Scotland - A Challenge for Devolution and the Renewal of Democracy

GREG LLOYD

Introduction

Quangos have been described as 'the fringe in the fog' of government (Jordan, 1994, 32). The sector is now a firmly established and integral part of the apparatus of governance and has replaced elected authorities as the natural home for government and the management of the public purpose (Skelcher, 1998). Quasi government raises questions about the accountability of government and the effectiveness of processes of governance. At the present time, the political agenda in the United Kingdom is dominated by a firm emphasis on the renewal of civil democracy, devolution, decentralisation, social inclusion and community empowerment (Giddens, 1998). Yet, it must follow that a continued or even increased reliance on quangos sits uncomfortably with this recasting of a middle ground between the conventional political debates associated with social democrats and neoliberals. This is certainly evident in Scotland at the present time, where the continued existence of quangos raises questions for the Scottish Parliament 'about the place of formal ministerial and parliamentary mechanisms in securing accountability for devolved functions' (Parry, 1999, 12). This chapter explores the nature of quasi-government in Scotland. It considers some of the principal issues raised for the Scottish Parliament and its interest in community planning as the basis of local governance in a devolved Scotland.

Origins and Nature of Quasi Government in Scotland

Quangos are defined as bodies, which have a role to play in the processes of national government, but are neither government departments nor, indeed, part of one. In practice, quangos operate to a greater or lesser extent at arms length from Ministers and government departments. It is important to stress, nonetheless, that although quangos may appear to operate independently of government, they remain ultimately accountable to Ministers (Cabinet Office, 1999). In practice, however, the lines of accountability may appear tenuous or non-existent, thereby contributing to a general suspicion of this layer of governance. In particular, quangos can attract considerable alarm and concern when viewed as a class of organisation which have considerable public significance yet remain largely outside democratic political activity (Skelcher, 1998, 1). This gives rise to what has been described as a 'governance gap' (Plummer, 1994). The associated institutional cleavage in the processes of governance is exacerbated further by the fact that, in recent times, the numbers of quangos have proliferated and increasingly mimic the structures and processes of private sector organisations (Jordan, 1994). This serves to accentuate the perceived distance between elected government and appointed bodies thereby contributing to a general distrust of this layer of governance with the potential to undermine the legitimacy of government itself.

In general, quangos have national, regional and local geographical jurisdictions. The emerging network of Regional Development Agencies being developed in England, for example, illustrate the national - regional level of involvement (Roberts and Lloyd, 1999). RDAs will be responsible for a spectrum of strategic functions, including economic development, social and physical regeneration, the provision of business support, the enhancing of skills and the promotion of sustainable development. It is sobering to reflect that these fundamental responsibilities for regional economies will be discharged and managed through quasi government rather than the directly elected bodies responsible to their local communities. The RDAs are primarily business-led, although they also include people with experience and expertise from local government and further and higher education, as well as trade unions and the voluntary sector. The intention is that they will bring greater coherence into national programmes by helping 'to integrate them regionally and locally' (Department of the Environment, Transport and the Regions, 1997, 16). The RDAs will establish a context to regional and local economic development in defined localities and will essentially create a quango led framework to established local authority and other initiatives and regeneration programmes. This will represent in short a powerful assertion and expression of the quango state. Increasingly, however, quasi government is emerging also as a powerful influence in terms of local governance. Davis (1996), for example, has documented the growth of government by appointment at the local level. In particular, he highlights the manner in which quango members directly confront the role of elected members. Indeed, there are a number of considerable differences between quango members and councillors that are of direct significance. For elected members, the principal differences relate to the representation of constituents in specific localities, the relationship to political activity, the wider range of service responsibilities involved, a liability for surcharge and the associated transparency of decision making (Skelcher and Davis, 1995). Furthermore, Hambleton (1997) has stressed the effects of quangos extending to the voluntary and community sectors in the processes of local governance.

Quasi government involves the transfer of power from the directly elected and relatively transparent system of central and local government to the closed world of quangos (Skelcher, 1998). The creation of arms length bodies is not new and since the turn of the century, quangos have played a significant role in national, regional and local governance. Indeed, Jordan (1994) interprets the recent history of quasi government as an example of the increasing uncertainty (along with inconsistency and disorder) in formal government itself in addressing increasingly complex policy arenas. The increase in the numbers of bodies that comprise quasi-government was initially viewed as a fourth sector in public administration, alongside the state, market and public corporations (Hague, Mackenzie and Barker, 1975). Indeed, it is interesting to note that public corporations themselves are now lumped together as part of quango-land (Weir and Beetham, 1998). This in itself is an interesting observation on the evolving state-market relationship throughout the former United Kingdom. The only real certainty here is the continual blurring of boundaries between public and private and the extent to which the private sector, business ethos is embedded in the design and execution of public policy. In the events leading up to devolution, for example, the quango issue was never far away from mainstream debates. In Wales, for example, a persistent theme in the lead up to the creation of the Assembly was that of a reform of the existing quangos, such as the Welsh Development Agency (Andrews, 1999). Its position, as the premier economic development body in Wales, was compounded by the changing political circumstances in which it operated, which had rendered it untenable in terms of the evolving democratic unaccountability (Morgan and Henderson, 1997). Similarly, in Scotland, the White Paper 'Scotland's Parliament', expressed concern at the extent to which public services were operated by unelected bodies and pledged to put arrangements in place to ensure greater transparency and accountability (Scottish Office, 1997).

The creation of quasi government to complement the work of formal government is essentially a response to the changing nature of government itself. Of necessity, in a rapidly changing social and economic context, policy and service delivery demands on central and local government have risen exponentially. By hiving off responsibilities to appropriate bodies, government has been able to maintain its focus on core policy matters. This apparent technical benefit of the quasi government sector was fostered in a sympathetic (neoliberal) political and ideological context. Throughout the 1980s and 1990s, for example, the constraints and limits placed on the public sector were accompanied by considerable institutional fragmentation and resulted in the remarkable creation of quangos. Further, the 'deconstruction of large public bureaucracies and their re-emergence as a multiplicity of smaller bodies in contractual and market type relationships to each other has been mirrored by the corralling of elected politicians into the role of strategic policy matters. The interpretation of their wishes, reconciliation of choices and purchasing and delivery of services is interpreted as a managerial task undertaken by an appointed board recruited principally for their business-oriented expertise. In the conventional wisdom, quangos provide an appealing and managerially efficient solution to the problems of governing a complex society' (Skelcher, 1998, 2). The general mistrust of government and local authorities in particular and the perceived need to focus on specific policy arenas, such as the inner city crisis, environmental protection and inward investment prompted a general acceptance and reliance on the quango state. Parry (1999, 18) has asserted that under 'the Conservative government, quangos, and especially local public spending bodies, became an alternative territorial system to the local authorities whose political affiliation moved completely away from the party'. As a result, funding was channelled into the quasi government state and away from the formal public sector.

Furthermore, a number of more specific motives of deploying quangos in the structures of government and the processes of governance may be identified. Parry (1999), for example, suggests their utility in legitimising an emerging public agenda; in distancing government from executive action by informed but impartial representatives; in the co-option of non-political elites into the governmental system to provide for a more informed administration and execution of policy; in enhancing expertise in decision making; and, in terms of facilitating central-local governmental relations. Indeed, Leicester and Mackay (1997) have identified the case for the continued deployment of quangos in a devolved Scotland. This would involve retaining functions at arms length from government when the execution of those specific functions are inappropriate for political intervention, when the functions are largely regulatory or operational, where the Minister in question has an appellate role against the decision of a quango, where functions involve commercial or entrepreneurial judgement, and where functions need to be discharged with the co-operation of a number of partners.

A Typology of Quasi Government in Scotland

There is considerable definitional diversity in the quango sectors. In reality, the constitution of quasi-government is confused, being characterised by a considerable blurring of organisational types. Within governments, quangos are more properly referred to as Non Departmental Public Bodies (NDPBs) but in practice and popular debate the term also includes other public bodies. These may not fall strictly within the formal classification of quangos but these tend to operate within a similar management framework. These other bodies include nationalised industries, public corporations and National Health Service bodies. Taken together, these constitute what may be described as quasi government. The boundaries of this world are not impermeable, however, and there are a host of other local bodies, partnerships and networks that exhibit similar operational characteristics to the quango sector. It is not the intention here to engage in a detailed discussion of the definitional debates concerning quasi government. The measurement of quasi government is relatively vague and reflects the different categories of organisation that may be included in the popular definition.

Whilst the generic concept of quangos captures a host of organisations working at arms length from government, considerable intellectual energy has been devoted to identifying a more sophisticated segmentation of the genre. Pliatsky (1980), for example, elucidated NDPBs which themselves are far from homogeneous and come in a variety of leadership, resourcing and jurisdiction forms. These then could be classified into executive, advisory and tribunal forms. This typology holds today as the basis for identifying the extent and nature of Scottish quasi government. Subsequently, however, a non-recognised layer of quango was identified; a category which tended to be excluded from official definitions (Weir and Hall, 1994; Painter, Isaac-Henry and Rouse, 1997). This wide array of bodies is confirmed by Davis and Skelcher (1997) who provide a bewildering review of the research conducted into the operation of quangos. Today, the organisation of the constituents of quasi government is still relatively primitive - the official categorisation of quangos considers executive quangos, advisory quangos, tribunals, public corporations and nationalised industries.

Secondly, the numbers of quangos and the composition of the quasi state are bewildering. Adopting the typology noted above, the evidence suggests for Scotland the following landscape of quango bodies.

Executive Quangos

These are the organisations with executive, administrative, commercial or regulatory functions. These bodies are typically engaged in a wide range of activities, often of a strategic nature, such as the Crofters Commission (with the remit to reorganise, develop and regulate crofting and to promote the interests of crofters) and Scottish Arts Council (which is charged with creating conditions in which the arts can flourish and be enjoyed throughout Scotland). The executive quangos include some of the more important interventionist mechanisms in the modern Scottish economy, including Scottish Enterprise and Highlands and Islands Enterprise, Scottish Environmental Protection Agency and Scottish Natural Heritage and Scottish Homes. These bodies cover the strategic functions of Scotland and engage with local government and communities in a variety of ways (Fairley and Lloyd, 1997; Mackay, 1995; Lloyd, 1999). The present executive quangos in Scotland are given in Table 8.1.

Advisory Quangos

These are the bodies set up by Ministers to advise them and their departments on specific matters. They are conventionally set up by administrative action and typically staffed by officials from within the sponsoring department. The advisory quangos currently operating in Scotland are shown in Table 8.2. Examples include the Ancient Monuments Board for Scotland which advises the Parliament on its functions under the appropriate legislation; the Scottish Childcare Board and the Scottish Valuation and Rating Council.

Tribunal Quangos

There are currently three tribunal quangos - the Children's Panel, the Horse Race Betting Levy Appeal Tribunal for Scotland and the Rent Assessment Panel for Scotland.

Public Corporations

The three public corporations in Scotland are the East of Scotland, North of Scotland and West of Scotland Water Authorities (PWAs). These are responsible for the provision of water and sewerage services within their defined geographical areas.   The PWAs are an outcome of local government reorganisation in Scotland which created a unitary system of general-purpose authorities but which separated out the responsibility for water and sewerage infrastructure (Black, 1994).

Nationalised Industries

These comprise Caledonian MacBrayne Limited, which is responsible for the operation of the main network of ferry services to the western isles of Scotland; Highlands and Islands Airports Ltd, responsible for the management and operation of airports in the region; and the Scottish Transport Group, a residual body dealing with matters relating to the sale of the Scottish bus companies which took place in 1990-1991.

National Health Service Bodies

These include a number of bodies concerned with the provision of medical, dental and mental health care services in Scotland and are shown in Table 8.3.

Principal Features of the Quasi Governmental State

An important feature of the quasi state is the scale of the quasi state in terms of expenditure and employment. Expenditure by the principal Scottish quangos is given for illustrative purposes in Table 8.4. This reveals a not insubstantial level of spend dedicated to strategic Scottish wide functions. Yet these fall outwith the direct democratic processes and control of government itself.   McQuaid (1999) has documented the principal employment details of the principal executive quangos in Scotland and has drawn attention to the uneven geographical distribution of that employment with a general concentration in Edinburgh. An illustration of this is given in Table 8.5. It is in this context that the quangos offer some potential for addressing the regional economic disparities in Scotland in terms of relative economic performance. As was suggested in the debates about the economic agendas for a Scottish Parliament, 'devolution undoubtedly does provide opportunities for influencing the Scottish economy, possibly significantly, the scale and even direction of effects depends on the particular contribution of policies pursued by the Scottish Parliament and the reactions of the Scottish people to them' (McGregor, Stevens, Swales and Yin, 1997, 208).

In this context, quangos could represent an important policy tool particularly with respect to the dispersal of employment. This, allied to those of the civil service, could form part of an innovative sub-regional policy within Scotland, as has been recognised in some areas (Highlands and Islands Enterprise, 1999). It would also re-assert a regional planning framework within Scotland which it could be argued was at least associated with the two-tier structure of local government which was replaced in 1996. It has been argued, for example, that the regional authorities

> Brought a vision and breadth of perception to the physical development of..[Scotland]...bringing together...economic land use and transportation needs within a coherent strategic framework......in a way which has reconciled the apparently irreconcilable and ensured a consensus amongst the wide range of public, private and voluntary interests' (Midwinter, 1995, xii).

Local government reorganisation served to erode that strategic element and resulted in its fragmentation into thirty-two unitary Councils. The importance of this in the context of the relationship between quangos and local government was highlighted in the decision making process associated with the Food Standards Agency in Scotland.

The Food Standards Bill was introduced in Parliament in June 1999 to protect public health in relation to food generally. The legislation was a response to a widespread loss of confidence in the public at large with respect to food safety and standards. This is not surprising given the unrelenting media attention on BSE, E-Coli and GM scares and debates. A broader European policy focus on food has served also to question the effectiveness of monitoring and standards in the UK. The principal focus of the Bill was to establish a quango - the Food Standards Agency - as the watchdog of a food safety policy initiative. This will be based in London but will have executive arms in Scotland, Wales and Northern Ireland. In Scotland, the Scottish Executive announced that the Agency would be based in either Dundee or Aberdeen. This led to a bidding process between the two cities to act as the host for the quango. Its significance being acknowledged in terms of institutional linkages, clustering of allied activities, prestige and local economic spin-offs. In the final outcome, the decision to locate the Scottish Food Standards Agency in Aberdeen was based on 'a wide range of factors, including transport links, accessibility to Ministers, MSPs and the UK Agency, links to research and scientific interests and relocation costs'. This process suggests that the full potential of quango dispersal has not yet been recognised by the Parliament and Executive.

In recent times, quangos have been subject to reconfiguration in line with prevailing political ideas. In the early 1990s, for example, a number of super-quangos were created in Scotland as a direct consequence of the prevailing political neoliberal ideology, a distrust of local government but also a perceived need to rationalise or streamline the then existing arrangements discharged by the quangos. The bodies included Scottish Enterprise, Highlands and Islands Enterprise, Scottish Homes, Scottish Environment Protection Agency and Scottish Natural Heritage. Scottish Enterprise and Highlands and Islands Enterprise were created to assume responsibility for the integrated delivery of economic and business development initiatives, the provision of training and measures to secure the improvement of the environment. These particular institutional arrangements replaced and extended an established tradition of regional development agencies in Scotland - the Scottish Development Agency and the Highlands and Islands Development Board together with the Training Agency (Danson, Fairley, Lloyd and Newlands, 1990). By this expedient, a rationalisation of agency was secured together with the integration of service within the quango state itself.

The Scottish Environment Protection Agency (SEPA) is another super executive quango with a complicated set of relationships with other bodies, including local authorities. In 1991, the Secretary of State for Scotland announced the Government's intention to create a new body to be responsible for the protection and management of the Scottish environment. It was intended to reflect the distinctive nature of established Scottish legal and administrative arrangements whilst meeting new environmental challenges. In introducing the proposed body, the Scottish Office stated that SEPA would assume responsibility for controlling defined forms of pollution in Scotland, including integrated pollution control, air quality and radioactive waste, the prevention and control of water pollution and the regulation of solid waste disposal. To this end, the Scottish Office stressed that SEPA 'will bring together the present regulatory bodies under an integrated structure... (and)...will take a strategic and pro-active approach to preventing and curbing pollution of the environment' (Scottish Office, 1991). In particular, SEPA was to streamline and strengthen the then existing administrative and organisational arrangements for environmental protection in Scotland which were characterised by considerable fragmentation of function and responsibility at central, regional and local levels of government. These arrangements for pollution control in Scotland exhibited considerable vertical and horizontal fragmentation of administrative responsibility and organisation. These included the Scottish Office which had (and retains) overall responsibility for environmental protection through promoting legislation and the publication of Circulars,

National Planning Policy Guidelines and Planning Advice Notes to establish the framework for environmental policy; Her Majesty's Industrial Pollution Inspectorate (HMIPI); the Health and Safety Executive; Scottish Natural Heritage; River Purification Authorities and local authorities which were responsible for waste disposal, local air pollution control and land use planning. The piecemeal nature of these arrangements together with their potential for duplication, waste and inefficiency had drawn critical attention from the Scottish Office (Lloyd, 1999). In similar fashion, the Natural Heritage (Scotland) Act 1991, for example, created Scottish Natural Heritage 'to secure the conservation and enhancement of and to foster the understanding and enjoyment of, the natural heritage of Scotland'. Scottish Natural Heritage (SNH) was formed through the merger of two organisations - the Nature Conservancy Council for Scotland and the Countryside Commission for Scotland to provide an integrated approach to environmental management and conservation of the natural heritage.

The super quangos involve a complex institutional structure in order to serve the diversity of the Scottish economy. Thus, Scottish Enterprise and Highlands and Islands Enterprise , for example, sub-contract delivery of the integrated training and development activities to a network Local Enterprise Companies (LECs). In practice, Scottish Enterprise and Highlands and Islands Enterprise provide strategic policy guidance and expert advice to the LECs - which are effectively mini-constituent quangos which operate within defined localities - on individual economic sectors; undertake major projects or research activities which extend beyond the areas of individual LECs; provide individual LECs with a range of central support services which will initially include administrative, accounting and property services; undertaking marketing and inward investment programmes for the areas in question; undertake major environmental improvement and land renewal programmes, in consultation with the LECs involved; and monitor the progress of the LECs in implementing their plans and achieving their objectives. The LECs are not statutory bodies but are private companies constituted under the Companies Act 1985 'to bring a direct knowledge and understanding of the needs and opportunities of the local economy to the delivery of the Government's enterprise, environment and training programme, and to engage the commitment, experience and entrepreneurial flair of senior members of the private sector' (Scottish Affairs Committee, 1995, v). The SEPA Board is appointed by the Secretary of State for Scotland and comprises a Chair, deputy Chair and ten members, including the Chief Executive. The Board appoints from its members the Chairs of the three Regional Boards and an Audit Committee, which ensures the effectiveness of SEPA's financial control systems. The SEPA Board has specific responsibility for establishing the overall strategic direction of SEPA; ensuring high standards of corporate governance; overseeing the delivery of planned results and ensuring that SEPA operates sound environmental policies in relation to its own operations. The three regional boards have general responsibilities include the development of business plans for the region, the generation and implementation of local initiatives for the environment and advising on applications which have significance for the environment. There are five Directorates: North Region, East Region, West Region, Environmental Strategy and Corporate Services.

Such a powerful concentration of influence has attracted concern, particularly with respect to the issues of accountability and the relationship between LECs and local government. The House of Commons Scottish Affairs Committee (1993 - 1995), for example, drew attention to the relationship of the LECs to the parent organisations and the relationships of the LECs to their communities for which they were responsible. The recommendations and conclusions of the Committee reflected this concern with the need to enhance accountability of the enterprise network particularly in the context of the relationships between LECs and their respective communities and localities. In this context, the Scottish Affairs Committee recommended, *inter alia*, that the geographical coverage of the LECs in the Scottish Enterprise network be reviewed to ensure coterminousity between their boundaries and those of local authorities. Further, the Committee wished to see local authorities and others in the public sector being fully involved with the development of the LECs' training strategy. The theme of accountability was pursued also through a recommendation that there be greater openness in the letting of contracts by LECs to local contractors - a policy already developed by Highlands and Islands Enterprise. Finally, the Committee drew attention to the importance of facilitating mandatory consultation by LECs of local authorities to replace the present discretionary arrangements (Fairley and Lloyd, 1995).

Towards Enlightenment?

There is the perceived governance gap which is perhaps the single most aspect of quangos that attracts the most critical attention from academics and other commentators. Plummer (1994), for example, has argued that there is great variety and inconsistency in the ways quangos are governed and, also, a remarkable lack of clarity about how they should be governed and to whom they are accountable. He has pointed to the remarkable discrepancies that constitute the governance gap, including those in established governance procedures, the recruitment of board members, the membership of individual quangos, the degree of independence enjoyed by the bodies concerned (which was not related to any apparent underlying rationale) and the regulatory regime and complaints procedures (Plummer, 1994). Yet, the question must still remain that these arguments for technical or administrative efficiency are at the expense of a trade off with accountability to specific interest, communities and individuals. Balancing this trade-off is delicate and difficult. It has been argued that from 'one perspective it is the weighing of public values and exercise of judgement - an essentially political activity - undertaken by a group of individuals appointed through a process of patronage and having no accountability to or legitimacy with citizens. From another perspective it is an efficient and politically astute way of governing and managing public services by drawing on the skills and experience of experts who, by virtue of being insulated from public view and party competition, are able to reach the best decisions for the community' (Skelcher, 1998, 5). Skelcher (1998) has suggested a number of ways available to reform quangos. These include procedures to secure the appointment of members on quango boards, through, for example, opening up the process through advertising which is now done in many of the Scottish quangos; establishing an Independent Appointments Commission and holding public hearings for key quango posts. Attention could be directed also to setting out clearer codes of conduct and provide for the regulation of conflict of interest. A more radical approach could be to strengthening accountability by transferring quangos to the elected sector, which is an unlikely proposition for the Parliment and could pose additional pressures on local authorities. Certainly, Skelcher (1988) advocates the greater involvement of the public in choices for quango boards in order to develop a dialogue with the community.

Since the mid 1990s, however, the public administration of quasi government has become more codified (Parry, 1999). This has included a system of reviews on a five yearly basis to oversee the financial affairs of individual quangos, the publication of codes of conduct for executive and advisory quangos to establish standards for those holding public office and to provide for the resolution of any conflicts of interest. There can be little doubt, however, that the principal impetus for this scrutiny of the broad behaviour of quasi government follows on from the deliberations of the Committee on Standards in Public Life. It reflects also the current political programme to effect a modernisation of government (Cabinet Office, 1999b). This sets out a long-term programme for the improvement in government by making it more inclusive and integrated, with strategic policy making and an emphasis on public service users rather than providers.   There can be little doubt that this focus of attention must embrace quasi-government.

In 1998, the Government published two documents which establish the policy context for quasi-government.   In 'Quangos - Opening the Doors', the Government restates its political commitment to keeping the numbers of quangos to a minimum.   Indeed, it asserts that quangos will only be set up where this is the most appropriate and cost-effective means of achieving the intended purpose (Cabinet Office, 1998a).   To this end, the government propose more open meetings and greater public access to the reporting of quangos.   It also advocates greater co-operation between quasi government and local authorities.   This is complemented by 'Quangos - Opening Up Public Appointments' which seeks to widen the field of candidates applying for public bodies and of improving representation among women and ethnic minorities (Cabinet Office, 1998b).   This latter initiative followed an earlier Scottish advertising campaign - 'Play a Part in Changing Scotland' - which was designed to broaden the social representation in Scottish public bodies.   In the context of a devolved Scottish Parliament, Leicester and Mackay (1997) argue that an appropriate way forward would be for the boundaries between the Parliament and Executive and individual quangos should be redrawn in the interests of promoting greater accountability. Perhaps the most telling argument is that a new clearly understood and democratised relationship if forged in order to legitimise the process of devolution itself. This is particularly the case as it would appear that the parliament will create new quangos as part of its own legislative programme, such as those associated with land tenurial reform.

Quangos and the Challenge for Community Planning

In Scotland, the process of community planning clamours for attention from the devolved Parliament. This touches directly on its relationship with local government and, by association, the quango state. The driving force was a concern with the emerging relations between the Scottish Parliament and the unitary local authorities (Alexander, 1997; McIntosh Commission, 1998).   In 1997, in anticipation of any possible conflicts over the still emerging institutional relationships, the (then) Secretary of State for Scotland and the Convention of Scottish Local Authorities (COSLA) established a joint body - the Community Planning Working Group - to consider the ways in which Scottish local authorities could engage in effective partnerships with other bodies to provide for and promote the economic, social and environmental well-being of the communities they serve.

The Community Planning Working Group (1998) view community planning as a process of public administration through which a Council exercises leadership and comes together with other organisations to plan, provide for, or promote the well being of communities they serve. The purpose of the community planning process is to present an informed view of the spectrum of interests of an area in terms of their perceived challenges and constraints, available opportunities, and the preferred priorities for action. In practical terms, the aims of community planning focus on the need to improve the service provided by Councils and their public sector partners to the public through closer, more co-ordinated working. Here, community planning is intended to provide a process through which Councils and their public sector partners, in consultation with the voluntary sector, business sector and the community, can agree both a strategic vision for the area and the action which each of the partners will take in pursuit of that vision. It is a process to help Councils and their public sector partners collectively identify the needs and views of individuals and communities and to assess how they can best be delivered and addressed (Community Planning Working Group, 1998). Indeed, it was asserted

> That there is at present a lack of structured overview across the various agencies about how they collectively could best promote the well being of their communities. The kaleidoscope of local and subject-specific plans and partnerships was not related to a consistent attempt to develop a sheared strategic vision for an area and a statement of common purpose in pursuit of that vision. Neither was it always clear who had the lead in developing this shared strategic vision (The Community Planning Working Group, 1998, 3).

Community planning involves also an outcome - a community plan - which would incorporate the priorities of the key actors in a locality and would assert a common agenda for action over a 5 - 10 year period and be subject to review. It would involve full consultation with individuals, communities and the private sector although it is clearly driven by the public sector community. In time, it is intended that the community planning process would work at more spatially disaggregated levels of community.

A defining feature of the community planning process and its intended outcome is that of the role of local authorities. These are expected to assert community leadership amongst the various bodies, groups and interests involved in a locality. This brings local authorities and the community planning process into a direct contradictory engagement with some of the principal (strategic) quangos. As a consequence of the fragmentation of the structures of government in the 1980s and early 1990s, the process of local governance in Scotland is heavily congested, in terms of the multifarious organisations, partnerships and agencies involved, and heavily contested, in terms of those bodies competing for scarce resources and policy influence. Many of these are quasi governmental bodies, such as the LEC networks of Scottish Enterprise and Highlands and Islands Enterprise. Other powerful agencies will be involved as key players in individual localities, including Scottish Homes, the European Structural Funds Partnerships, Scottish Natural Heritage, Scottish Environmental Protection Agency, and the Public Water Authorities, Health Boards and joint Police Boards. This will involve local authorities having to operate in an assertive (reticulist) manner in such a way as to identify common perceptions of prevailing and anticipated challenges for the locality, broker the reconciliation of local strategies, policies and resource allocation programmes together with ensuring the stability of inter-organisational relations in the emerging spatial partnerships. Local authorities will have to operate as the fulcrum to these processes and this will demand considerable negotiating skills, resources and diplomacy to bring the process to an effective outcome (Lloyd and Illsley, 1999).

The evidence reveals the role played by quangos in this (allegedly) democratising process. In order to facilitate the community planning initiative, Pathfinder Councils were nominated - Highland Council, City of Edinburgh Council, South Lanarkshire Council, Perth and Kinross Council and Stirling Council - to produce pathfinder community plans. These reveal that each embraced the broad philosophy of community planning in very different ways in terms of process and outcome. In general, the pathfinders engaged in a broad consensus building approach to devising an appropriate strategic vision for their specific arenas. Whilst there were variations in approach, all tended to rely on the key quangos in their areas. In Perth and Kinross, for example, the Council identified a number of key agencies (Scottish Enterprise Tayside, Scottish Homes, Tayside Health Board, Tayside Police and Perth College) as the basis of its partnership for 'long term vision, common understanding, sharing of information, co-ordination of activities and agreement on priorities for the future'. To secure implementation and to transmit the process into the wider community, the Perth and Kinross plan sets out a Strategy Group to monitor progress and to involve citizens and communities in planning the delivery of services throughout the area. This suggests that the governance gap associated with quangos is legitimised at the very core of the community planning initiative. It contradicts the nomenclature of the exercise and could undermine the leadership role of local authorities.

Conclusions

Times are a changing. The structures of public administration and government, the processes of governance and the delicate inter-relations and power balances involved in service delivery, policy design and implementation and resource allocation are being reconfigured under the banner of devolution. There is considerable stress on the greater openness of decision making and policy execution and the empowerment of communities. In abstract or national (or political) terms, these ideas are important yet need to be grounded. It has been argued, for example, that 'public policy is typically framed as if place and space were irrelevant. There are many arguments in the contemporary period to support the neglect of place and space. The dynamics of our economies these days seem to have floated free of locations and borders. Strategies to deregulate markets pay little attention to spatial consequences. New forms of regulation, even of environmental externalities, focus on the site or the firm, not the site and the firm in their local setting' (Healey, 1998, 2). It is at this level - the neighbourhood, the community and the region - that the intended change in responsibility and influence will be experienced the most acutely. It is at this level that the potential governance gap and the inefficiencies associated with inter quango conflicts (or shortfalls) will be most focused. The greater the national political hype, the greater the local contradiction and failed expectations. Quangos are powerful bodies. They discharge strategic functions and, as an outcome of a historical reliance on them, are deeply structurally embedded in many local and regional economies, localities and communities. Yet, quangos are unelected and this position must be remedied to conform to the intentions of the prevailing political process of decentralisation and devolution. This will be a major challenge for the Parliament and may prove to be its defining contribution to a new Scotland.

Table 8.1 Executive Quangos in Scotland

Accounts Commission for Scotland
Hannah Research Institute
Modern Research Institute
Maculae Land Use Research Institute
Rowe Research Institute
Scottish Crop Research Institute
Crofters Commission
Deer Commission
National Board for Nursing Midwifery and Health Visiting
National Galleries of Scotland
National Library of Scotland
National Museum of Scotland
Parole Board for Scotland
Royal Boating Garden
Royal Commission on the Ancient and Historical Monuments of Scotland
Scottish Agricultural Wages Board
Scottish Arts Council
Scottish Children's Reporter Administration
Scottish Community Education Council
Scottish Convincing and Executor Services Board
Scottish Council for Educational Technology
Scottish Further Education Unit
Scottish Homes
Scottish Hospitals Endowments Research Trust
Scottish Law Commission
Scottish Legal Aid Board
Scottish Medical Practices Committee
Scottish Natural Heritage
Scottish Screen
Scottish Qualifications Authority
Scottish Sports Council
Scottish Tourist Board
Scottish Water and Sewerage Customer's Council

Table 8.2 Advisory Quangos

Advisory Committee on Dental Establishments
Advisory Committee on Sites of Special Scientific Interest
Ancient Monuments Board
Building Standards Advisory Committee
Central Advisory Committee on Justices of the Peace
General Teaching Council for Scotland
Health Appointments Advisory Committee
Hill Farming Advisory Committee (Scotland)
Historic Buildings Council
Justices of the Peace Advisory Committees
Local Government Boundary Commission for Scotland
Local Government Property Commission
Parliamentary Boundary Commission for Scotland
Post Qualification Education Board for Health Service Pharmacists in Scotland
Royal Fine Art Commission for Scotland
Scottish Advisory Committee on Drug Misuse
Scottish Advisory Committee on the Medical Workforce
Scottish Consultative Council on the Curriculum
Scottish Crime Prevention Panel
Scottish Industrial Development Advisory Board
Scottish Records Advisory Council
Scottish Standing Committee for the calculation of residual values of fertilisers and feeding stuffs
Scottish Studentship Selection Committee
Scottish Valuation and Rating Council
Advisory Group on Sustainable Development
Advisory Committee on Scotland's Travelling People
Secretary of State (Electricity) for Scotland's Fisheries Committee

Table 8.3 National Health Service Bodies

Common Services Agency for the NHS
Health Education Board for Scotland
Mental Health Commission
Scottish Council for Postgraduate Medical and Dental Education
Scottish Hospitals Trust
State Hospitals Board

Table 8.4 Expenditure by Principal Quangos in Scotland, 1998

Principal Bodies	Expenditure £m
East of Scotland Water	201
North of Scotland Water	128
West of Scotland Water	247
Scottish Enterprise	462
Highlands and Islands Enterprise	77
Scottish Environment Protection Agency	28
Scottish Further Education Funding Council	552
Scottish Homes	399
Scottish Tourist Board	25
Scottish Natural Heritage	37
NHS Trusts (47)	3.2 ban
Health Boards (15)	4.3 ban

Table 8.5 Quango Employment per 10,000 Population, 1996

City	NDPB: employment/10,000 residents	NDPB plus PWA: Employment /10,000 residents
Glasgow	24	45
Edinb.	59	75
Aberdeen	35	54
Dundee	33	58
Stirling	31	58
Inverness	51	90
Scotland	16	29

Source: McQuaid (1999, 153)

References

Alexander, A. (1997), 'Scotland's Parliament and Scottish Local Government: Conditions for a stable relationship', *Scottish Affairs* 19, 22 - 28.

Andrews, L. (1999), *Wales Says Yes. The Inside story of the Yes for Wales Referendum Campaign*, Bridgend, Seren.

Black, S. (1994) 'What's happening to water and sewerage services in Scotland?', *Scottish Affairs*, 6, 25-34.

Cabinet Office (1998b), *Quangos - Opening the Doors*, London, June.

Cabinet Office (1998b), *Quangos - Opening up Public Appointments*. London, June.

Cabinet Office (1999a), *Public Bodies 1998*. Cabinet Office, London.

Cabinet Office (1996b), *Modernising Government*, A White Paper, London.

Community Planning Working Group (1998), *Report of the Community Planning Group*, Scottish Office, Edinburgh, July.

Danson, M. Fairley, J. Lloyd, M.G. and Newlands, D. (1990), 'Scottish Enterprise: An evolving approach to integrated economic development in Scotland, in Brown, A. and Parry, R. (eds), *The Scottish Government Yearbook 1990*, University of Edinburgh, Edinburgh, 168-194.

Davis, H. (1996), *Quangos and local government. A changing world*, Frank Cass, London.

Department of the Environment, Transport and the Regions (1997) *Regional Development Agencies*. Discussion Paper, London, August.

Fairley, J and MG, Lloyd. (1995), 'Scottish Enterprise and Highlands and Islands Enterprise - a preliminary assessment and some critical questions for the future', *Regional Studies*

Giddens, A. (1998), *The Third Way. The Renewal of Social Democracy*, Polity Press, Cambridge.

Hague, D.C. Mackenzie W.J.M. and Barker A (1975), *Public Policy and Private Interests. The Institutions of Compromise*. London, Macmillan.

Hebbert, M. (1992) Environmental foundation for a new kind of environmental planning. *Town and Country Planning*, June, 166-168.

Healey, P. (1997), 'The revival of strategic spatial planning in Europe', in Healey, P. Khakee, A., Motte, A. and Needham, B. (eds), *Making Strategic Spatial Plans*, UCL Press, London, 3 - 19.

Highlands and Islands Enterprise (1999), *Dispersal of civil service jobs to the Highlands and Islands*, HIE, Inverness.

Jordan, G. (1994), *The British Administrative System. Principles versus Practice*. London, Routledge.

Leicester, G. and Mackay, P. (1998), *Holistic Government. Options for a Devolved Scotland*, Scottish Council Foundation, Edinburgh.

Lloyd, M.G. (1999), Scottish Environmental Protection Agency: Making Sense of a Fragmenting Environment, *Scottish Affairs*, 29, 28-42.

Lloyd, M.G. and Illsley, B.M. (1999), 'An idea for its time? Community planning and reticulism', in Scotland. *Regional Studies* 33(2), 181-184.

Mackay, D (1995), *Scotland's Rural Land Use Agencies*, Aberdeen, Scottish Cultural Press.

McIntosh Commission (1998), *Local government and the Scottish Parliament*, Edinburgh.

McGregor, P.G. Stevens, J. Swales, JK. and Yin, YP. (1997), 'Some simple macroeconomics of Scottish devolution, in Danson, M.D. Hills, and Lloyd, M.G. (eds), *Regional Governance and Economic Development*. European Research in Regional Science 7. Pion, London, 187-209.

McQuaid, R. (1999), 'The local economic impact of the Scottish Parliament', in McCarthy, J. and Newlands, D. (eds), *Governing Scotland: Problems and Prospects, The economic impact of the Scottish Parliament*, Ashgate, 149-166, Aldershot.

Midwinter, A. (1996), *Local government in Scotland. Reform or decline?* Macmillan, London,.

Morgan, K. and Henderson, D. (1997), 'The Fallible Servant: Evaluating the Welsh Development Agency', in Macdonald, R. and Thomas., A (eds), *Nationality and Planning in Scotland and Wales*, University of Wales Press, 77-79, Cardiff.

Painter, C., Isaac-Henry, K. and Rouse, J. (1997), 'Local authorities and non-elected agencies. Strategic responses and organisational networks', *Public Administration* 75 (2), 225-246.

Parry, R. (1999), 'Quangos and the Structure of the public sector in Scotland', *Scottish Affairs*, 29, 12-27.

Pliatzky, L. (1980), *Report on Non-Departmental Bodies*, Cm 7797 , London.

Plummer, C. (1994), *The governance gap. Quangos and accountability*, Joseph Rowntree Foundation, London.

Roberts, P.W. and Lloyd, M.G.(1999), *Realising Regional Potential*, BURA.

Scottish Affairs Committee (1995), *The Operation of the Enterprise Agencies and the LECs*, Volume 1, HMSO, London .

Scottish Office (1991), *Scotland to have its own environment protection agency*, Press Release, September, Edinburgh.

Scottish Office (1992), *Improving Scotland's Environment - The Way Forward*, A Consultation Paper. Edinburgh.

Scottish Office (1995*), Statutory guidance to the Scottish Environmental Protection Agency*, Draft Consultation Paper, Edinburgh, December.

Scottish Office (1997), *Scotland's Parliament*, Edinburgh, Cm 3658, July.

Sinclair, D. (1997), 'Local Government and a Scottish Parliament'. *Scottish Affairs* 19, 14 - 21.

Skelcher, C. (1998), *The Appointed State. Quasi-governmental Organisations and Democracy*, Open University Press, Buckingham.

Skelcher, C. and Davis H. (1995), *Opening the board room door. The membership of local appointed bodies*, Joseph Rowntree Foundation, London.

Stewart, M. (1998), Partnership, leadership and competition in urban policy, in Oatley N. (ed), *Cities, Economic Competition and Urban Policy*, PCP, 77-90, London.

Weir, S. and Hall, W. (1994), *EGO-Trip: Extra-Governmental Organisations in the United Kingdom and their accountability* . London, Charter 88

Weir, S. and Beetham, D. (1998) *Political power and democratic control in Britain*, Routledge, London.

PART IV
INTER-GOVERNMENTAL RELATIONS: IRELAND - EU - GLOBAL

9   00 New Relations Between Scotland and Ireland

GERARD MURRAY

Introduction: The Northern Ireland Peace Process

The people of Ireland, Britain and the international community were somewhat amazed that the two polarised communities in Northern Ireland, namely the Protestant/Catholic, unionist/nationalist, loyalist/republican communities, could reach an historic compromise by signing the Good Friday Agreement on the 10 April 1998. The converse occurred on the 15 July 1999 when the familiar scenes of intransigence, enmity and lack of trust resurfaced at the failure of the Ulster Unionist Party to form a powersharing executive with two Sinn Fein prospective ministers. Unionists were judged by the international community to be perpetrators of a policy which locked the door to devolved government in Northern Ireland. The thorny issue of decommissioning of weapons has always been at the crux of the Northern Ireland Peace Process. Martin McGuinness of Sinn Fein, in reaction to unionist opposition to form a powersharing administration with his party, claimed on the BBC programme *On the Record*, 'This is not about decommissioning. It is about rejectionist unionists not wanting to give equality to nationalists; not wanting Catholics in government; north-south implementation bodies; disbandment of the RUC.' [1]

Despite this chilling political diagnosis from a leading republican, even against all the odds - the use of semantics, and political spindoctoring - the second Mitchell review moved Northern Ireland in December 1999 to have a devolved powersharing administration, with two Sinn Fein Ministers holding the Education and Health portfolios. The Ulster Unionist leader, David Trimble, entered into a partnership government with republicans with a tacit understanding emanating from the Mitchell Review that a start to decommissioning would take place by the beginning of February 2000. The high moral ground that republicans enjoyed in July 1999 now shifted to the unionists. The subsequent collapse of the powersharing executive was a direct consequence of the IRA failure to start the process of decommissioning and indeed undermined McGuinness' July denouncement of unionists.

Following the decision of Northern Ireland Secretary of State, Peter Mandelson, to suspend the devolved institutions in the province from midnight on the 11 February 2000 political uncertainty over the future of the Peace Process has returned to Northern Ireland. However, in a dramatic turn around of events the IRA issued an unexpected statement on the 6th May 2000 stating: 'the IRA leadership will initiate a process that will completely and verifiably put arms beyond use' (RM Distribution, 6 May 2000). The onus for the restoration of devolved institutions in Northern Ireland now hinges on a meeting of the Ulster Unionist Council which takes place on the 27th May 2000. David Trimble requires a majority vote at this historic gathering to offset the demise of the Good Friday Agreement (see end of chapter). Republicans also could renege their commitment to the Belfast accord if concessions are made to unionists in relation to the Royal Ulster Constabulary (RUC). Trimble realistically acknowledges the alternative to the Good Friday agreement is direct rule with a stronger Dublin input leaving unionists in the abyss with 'no say at all'.[2]

Council of the Isles

Tom Hennessey, a former Ulster Unionist Party (UUP) negotiator at the multi-party talks leading to the signing of the Good Friday Agreement on 10 April 1998 has argued that the UUP viewed the north-south bodies during the negotiations as part of the Council of the Isles.[3] This position has been in stark contrast to the view of the SDLP and Sinn Fein parties who wanted north-south bodies as separate and freestanding from any other institutions. The UUP viewed all the devolved legislators along with Westminster and the Dail as separate entities - but all part of the greater whole. During the multi-party talks process, the UUP argued for the British-Irish Council (BIC) or the Council of the Isles as an umbrella organisation under which north-south bodies would exist. For unionists the east-west relationship within the British Isles between the United Kingdom of Great Britain and Northern Ireland and the Republic of Ireland, is far more important than the north-south relationship.

The UUP view the BIC as contained within the Good Friday Agreement as their idea. Brendan O'Leary notes that despite unionists originally desiring, 'any North-South Ministerial Council to be subordinate to a British-Irish, or East-West Council. This has not happened.'[4] However, the BIC is not only attributed to Ulster Unionists in Northern

Ireland; but includes the Ulster Democratic Party (UDP) who suggested the formation of a Council of the Isles in their 1986 document *Common Sense*. The UDP viewed the Council of the Isles as an organisation to embrace economic, cultural, technological, standards and EC matters. In 1996 the SNP commissioned an enquiry into how Scotland could move from being part of the United Kingdom to becoming independent within Europe. Paragraph 68 of the report stated: '...that following independence a body analogous to the Nordic Council to be established for regional co-operation among the states of the British Isles, including the Republic of Ireland. Membership would be open not only to the independent states of the British Isles but also to autonomus islands such as the Isle of Man and the Bailiwicks of Jersey and Guernsey'.[5] The Council of the Isles is a Scottish Nationalist Party concept that envisages an 'Association of British States' after independence. The role of the association would be advisory/parliamentary and at civil service level not far off what is contained in the Belfast agreement. In other words Alex Salmond's standpoint on a post-independent position. The social union would continue and develop with England and Wales on social and economic matters. Gerry Hassan has also pointed out, 'Separatism ... does not exist even in the politics of Scottish independence'.[6]

The Republic of Ireland 'Celtic Tiger-economy' built up only over the last decade with its strong affiliation to the European Union has broken the stereotypical notion of what it is to be Irish. There is a strong sense of pride in being Irish, worldwide, but that is particularly relevant only if you come from the Republic. To be Irish in Northern Ireland has mixed connotations. There, the stereotype runs as follows: if you are a nationalist in Northern Ireland your Irishness is supposed to be similar to that of your counterpart in the Republic. In reality partition largely worked for the Republic but as O'Halloran has argued, southern society traditionally grouped northern people - both unionists and nationalists - all as one.[7] There was an understandable indifference towards Northern Ireland from the Republic, just as has been the case in Scotland and the rest of the United Kingdom.

From the Unionist perspective it was, and is quite acceptable to be Irish in a British context in the same way that Scots can be Scottish and British, and the people from Wales can be Welsh and also British. However, as devolution unfolds *Britishness* in terms of cultural, religious and political identification within each of the regions of the United Kingdom will be redefined. As part of this redefinition of identity; a strong

East-West institutional body involving representation from the Scottish Parliament could help give security to unionists. Scots would also find a Council of the Isles useful as a forum within which to consult their counterparts in Northern Ireland and the Republic to determine and refine their identity within a changing Scottish, British and European context.

British Irish Council

Scotland and Wales have recently established devolved institutions in each of their respective jurisdictions while the UK Government programme of constitutional reform leaves open the possibility for further decentralisation to the English regions. As part of these revolutionary changes: a new British-Irish Council was created through the Good Friday Agreement in April 1998. The BIC is commonly viewed as a conciliatory offering to unionists in Northern Ireland to alleviate their apprehensions over cross-border bodies as outlined in the Good Friday Agreement. I would argue that this has been the case set against the background of the Framework Documents of 1995. In the Framework Documents *totality of relationships* refers particularly to the north-south dimension. A couple of paragraphs are supposed to be on the east-west relationship; but in effect only five lines relate to east-west, and the rest is specifically on north-south issues.

The first meeting of the BIC on the 17 December 1999 brought together elected heads of the British and Irish governments as well as those of the devolved bodies in Edinburgh, Belfast and Cardiff, and the Channel Islands and the Isle of Man. David Trimble said at the inaugural meeting, 'Our view of the world must not be focused purely on Whitehall. What we must do is respect diversity throughout the British Isles.' His comments demonstrate his potential as First Minister in Northern Ireland to develop co-operation with the other devolved institutions throughout the United Kingdom. Lessons from the development of European institutions could also be used for championing the principles of the BIC.

John Hume has argued that we are living in a postnationalist era. He views the European Union as the greatest example of co-operation between countries who had traditionally been at war. European countries had come together as relatively recent institutions and formed a model for co-operation between diverse states. Hume has pointed to the fact that the political institutions of the European Union are a clear demonstration of institutional capacities to accommodate peoples of ethnic and racial diversity. Hume has said, '...fifty years after World War II, as a result of an agreed process, [European countries] have now been able to create one parliament to represent them ... They have a unity in diversity'. He has also pointed out, 'The answer to difference is to respect it, to create institutions which respect differences and gives victory to no one.'[8] One of the realities of Northern Ireland is the level of parochialism that exists there. The Northern Ireland conflict is based on religious and political difference. An institution such as the BIC can provide a mechanism for Northern Ireland to learn to live with difference through formal engagement with the rest of the peoples of the British Isles.

However, the degree of co-operation within the BIC is initially only at the stage of sharing of policy ideas between the new devolved institutions; joint action programmes are not yet being considered. The Scottish First Minister, Donald Dewar confirmed this position during a special debate on the BIC in the Scottish Parliament on the 2 February 2000. He told MSPs,

> I occasionally see rather over-ambitious definitions of the possible future role of the British-Irish Council, so it is worth making the point that no one envisages that it is - or will be in the immediate future other than a place for the exchange of ideas, building contacts and learning from one another.[9]

Mr Dewar explained further,

> The idea is not that the Council will be in permanent session, but that there will be occasional gatherings of the full Council, in between which working groups will prepare papers, explore possibilities and conduct conversations in a civilised and, I hope productive way on the chosen topics.[10]

The SNP point to Section Three, Paragraph 10 of the Good Friday Agreement in relation to the British-Irish Council which states:

> In addition to the structures provided for under this agreement, it will be open to two or more members to develop bilateral or multilateral arrangements between them. Such arrangements could include, subject to the agreement of members concerned; mechanisms to enable consultation; co-operation and joint decision-making on matters of mutual interest; and mechanisms to implement any joint decisions they may reach. These arrangements will not require the prior approval of the BIC as a whole and will operate independently of it.

In effect the SNP argue that if the Scottish Parliament and the Dail want to forge an agreement around particular devolved issues, Westminster is actually irrelevant.[11] Additionally in the debate over the BIC Alex Salmond pointed out,

> The First Minister is right to say that the Council is not a federal body making decisions.... However, where agreement can be reached on a multilateral or bilateral aspect as provided for in Strand 3 of the agreement, we should expect action to follow. The Council is not just a talking shop, but an institution where action will follow based on mutual or multilateral agreement.[12]

The Celtic Tiger

Before 1990 the Irish Republic was renowned as an underdeveloped country largely based on an agrarian economy. The unionist economic argument was traditionally that Northern Ireland under the auspices of the United Kingdom was better off than being part of a united Ireland. John Whyte has noted that large-scale emigration from the Republic offset the country's 'collapse'. He has cited Paisley et el who claimed that to amalgamate with the Republic 'would be to join economic hopelessness and a huge debt'.[13] In sharp contrast to this negative prognosis on the 17 December 1999, the Irish President, Mary McAleese, addressing the joint Houses of the Oireachtas said, 'Today's Ireland is a first-world country with a third-world memory'.[14] She wondered who could have predicted a reversal in the trend of emigration and Ireland's ill reputed economy a decade ago?

There is understandable interest by Scotland in the Republic's flourishing economy and its strong position within the European community. The Irish Consulate in Scotland has been working on the possibility of establishing a Scottish-Irish business forum. Figures provided by the Irish Consulate in Scotland show the UK is Ireland's largest trading partner accounting for 22 per cent of Ireland's exports in 1998. The Irish market is also very important to Britain; Ireland is the fifth largest export customer to the UK. One has to assume similar patterns between Scotland and the Republic. A survey of Scottish Manufacturing and Exports 1996-97 by the Scottish Council, Development and Industry, shows that out of 40 top markets for Scottish manufactured exports in 1996 Ireland (Republic) came twelfth with £353 million worth of exports. However overall 90% of the Republic of Ireland's exports in 1949 went to the UK, and in 1997 this was down to 24%.[15]

The Scottish National Party regularly draws comparisons with the Irish Republic and the country's recent success within the European Union.[16] Alex Salmond took the opportunity at the Humbert Summer Schools in Ballina, County Mayo in 1996 and 1998 to discourse on *Northern Ireland and the Scottish Question.* On the 23 November 1996 Mr Salmond stated, 'Ireland is doing so much with so little, while we are doing so little with so much.'[17]

Scottish-Irish Co-operation

Paragraph 5 of the recent government legislation to suspend the Belfast assembly meant that the functions conferred by section 52/3 of the Northern Ireland Act 1998 to the north-south bodies and the British-Irish Council are not to be exercised. The relatively recent presence of an Irish Consulate in Scotland is a clear indication that with or without a fully operational BIC, the Republic of Ireland will maintain, at a minimum, bilateral contact with the Scottish Executive with the objective of developing areas of co-operation at social, economic and cultural levels. The fact that Scotland has an Executive means that the Irish government has a political engagement in Scotland. Therefore links between Scotland and the Irish Republic can be independent of the British Irish Council. Without the formal implementation of the BIC Northern Ireland would lose significant economic and political connections with Scotland.[18]

The absence of devolved institutions in Northern Ireland means that Scotland has stronger links with the Irish Republic than Northern Ireland. Traditionally, students coming to study in Scotland from Northern Ireland are from the Protestant community. Eight % of the student population of Dundee University, representing 878 students in total, are from Northern Ireland.[19] There were 5,591[20] full-time undergraduates from Northern Ireland studying at Scottish universities during the 1998/99 academic year in contrast to only 1,576[21] from the Republic. Now, as the Irish Consulate in Edinburgh strengthens links between Scotland and the Irish Republic, more students are likely to take up Scottish University places, irrespective of the Good Friday Agreement.

Currently Northern Ireland is out of sequence with the rest of the unfolding devolution settlement within the United Kingdom. Traditionally where political, religious and personal links were much more developed between Northern Ireland and Scotland, the converse is taking place irrespective of the collapse of the Northern Ireland Peace Process or of the implementation of the BIC. It must be noted that Taosieach, Bertie Ahern, said at the inaugural meeting of the BIC, 'We must accept each other's circumstances, and accept - as a positive rather than a negative factor - that not all of us will wish to engage in all of the activities taking place in the framework of the Council.' Perhaps he indicates a somewhat lukewarm response to the BIC on the part of the Irish government. After all the Irish government wish to focus on diplomatic relations through their Consulate in Edinburgh.

If the Good Friday Agreement is not fully implemented alternative east-west mechanisms need to be put in place. Strand three, east-west relations as outlined in the Good Friday Agreement are not a totally novel concept, but can be traced back to the first major Anglo-Irish summit in Dublin between then Taoiseach, Charles Haughey and Prime Minister Margaret Thatcher on 8 December 1980. The meeting covered a range of international issues including the future development of the European Community; the European budget; Common Agricultural Policy; European Monetary System and other matters of concern to both countries. In a joint communiqué released from the leaders on the 8 December 1980 it stated that since their meeting of 21 May 1980 there had been 'useful exchanges at Ministerial and official level...leading to new and closer co-operation in energy, transport and security'. Paragraph 4 of the communiqué stated,

> The Taoiseach and the Prime Minister agreed that the economic, social and political interests of the people of the United Kingdom and Northern Ireland and the Republic are inextricably linked, but that the full development of these links has been put under strain by division and dissent in Northern Ireland. In that context they accepted the need to bring forward policies and proposals to achieve peace, reconciliation and stability; and to improve relations between the peoples of the two countries.

For this purpose they commissioned joint studies covering a range of issues including possible new institutional structures, security matters, economic co-operation and measures to encourage mutual understanding. Similarly, joint studies between Scotland, Northern Ireland and the Irish Republic could take place covering a range of topics. In relation to Northern Ireland these could include fisheries, transport and tourism.[22]

At the inaugural meeting of the BIC on the 17 December 1999 the Irish government agreed to work on a range of issues: drugs; the Scottish Executive and the Welsh Cabinet; social exclusion; the Northern Ireland Executive; transport; British government; environment; the knowledge economy; and links with offshore economies such as Jersey. These remain identifiable areas where practical joint studies could take place between Scotland and Ireland. Currently there is interest from the Scottish Executive in Ireland's strategy to combat drug dealers. In particular the Scots are interested in the powers of the Republic of Ireland Criminal Assets Bureau to seize money and property from suspected drug dealers. The environment, and the issue of Sellafield is a longstanding concern to environmental bodies protecting the East Coast of Ireland. The Irish government is concerned at the threat Sellafield poses, particularly around the Louth area. The Scottish Parliament will almost certainly take a similar view at the threat of Sellafield to the West Coast of Scotland.

The ferry service between Campbeltown and Ballycastle are outward signs of the important links between Northern Ireland and Scotland. In the discussions surrounding the BIC Dr Winnie Ewing of the SNP outlined how 'all the Irish and Scottish members of the European Parliament co-operated in securing support for the Ballycastle-Campbeltown ferry'.[23] The Research Institute for Irish and Scottish Studies at Aberdeen University and the parallel development of a department at University College Dublin are also outward signs of the new education links betyween Scotland and Ireland.

The Scottish-Irish Business Forum was launched in Edinburgh on the 11 March 1999 to develop links between Scotland and Ireland. It is also intended that the new body would promote joint Scottish and Irish ventures in the European market. The Convention of Scottish Local Authorities (COSLA) is attempting to create a new forum echoing the framework of the BIC. The idea is to have local government representation from England, Wales, Northern Ireland and the Republic of Ireland under the proposed body meeting to discuss issues of mutual concern.[24]

The Glencree Centre for Reconciliation in County Wicklow set up a New Horizons Programme just over two years ago and perhaps reflects a microcosm of the British Irish Council. During its first year over 1,000 participants from Bray outside Dublin, the Waterside in Derry/Londonderry, Govan in Glasgow and Wester Hailes area of Edinburgh took part in a series of community relations and community development projects. A major project involving an examination of how each of the communities addresses the drug problem took place in autumn 1999.

The Scottish Parliament: Lessons for Irish Society

One particular kind of sectarianism is demonstrated in Scotland and Northern Ireland through sport.   Donald Findlay, QC, lost his vice-chairmanship of Ranger's Football Club after being caught on video singing anti-Catholic songs at a post-match celebration.   His insults did little to persuade Scottish Catholics that sectarianism is a thing of the past in Scotland.   Joyce McMillan a leading Scottish arts commentator stated:

> ...the kind of sectarian feeling displayed by those football supporters is not so much the tip of some huge iceberg of hatred, discrimination, and implacable political disagreement about Scotland's future, but the last hurrah of a kind of stupid racism against the great immigration of the nineteenth-century that Scotland should have left behind long ago; but which - alarmed by the example of Northern Ireland, and scared of rousing old demons - we have rather preferred not to mention.[25]

Scottish composer James MacMillan also contributed to the sectarian debate by making the headlines during 1999 arguing that in all spheres of life Catholics are not full citizens in Scotland.[26]   The weakness of his claims lay in the absence of substantive empirical evidence.   However, Steve Bruce, Professor of Sociology at Aberdeen University has pointed out that comparison between Catholics in Northern Ireland and Scotland is markedly different.   In Scotland there are relatively few areas residentially segregated compared to 90% in Northern Ireland.   In Scotland half the Catholic population marry non-Catholics compared to only 5% in Northern Ireland. In particular Bruce argues that 'in Scotland all the social surveys of the last two decades show no discernible difference in the socio-economic status of Catholics and non-Catholics'.[27] On the contrary there is evidence to suggest that Scotland's new parliament is demonstrating that it does not tolerate sectarianism in Scottish society. The SNP move to get the Scottish Parliament to discuss changing the Act of Settlement, which forbids Catholics from ascending the throne, demonstrates two important factors. The significant factor is the extent to which the origins of the debate comes from a party traditionally linked with the non-Catholic community in Scotland shows the extent to which the Scottish Parliament is attempting to make a pluralist and inclusive society for all its citizens. A parallel Northern Ireland situation would be akin to the Ulster Unionist Party in Northern Ireland taking the lead by practical confidence-building measures to integrate the Province's Catholic community into the fabric of society.

Changing the Act of Settlement was debated in the Scottish Parliament on the 16 December 1999. Although the Parliament has not the legislative power to change the law it is nevertheless setting a precedent for

Westminster to take on board the views of the Scottish Parliament.[28] The original motion by Mike Russell of the SNP calling for the debate was backed by 77 of the 129 MSPs, representing cross-party support.[29] This scale of support demonstrates that the Scottish Parliament is taking the lead by signalling to London and Belfast that discrimination in any form has no place in Scottish society. Such public demonstrations for tolerance and pluralism in Scotland by the Scottish Parliament are commendable and are an example to Northern Ireland in the absence of a devolved administration in the Province.

The Scottish Executive is showing signs of courage in implementing pluralist policies despite opposition within conservative and discriminatory sections of Scottish society. The Scottish Parliament is pressing ahead with measures to repeal Section 28 of the Ethical Standards in Public Life Bill to allow homosexuality to be discussed in schools. This was despite the unhelpful comments from Catholic Cardinal Winning and millionaire tycoon Brian Souter who have contributed unwelcome homophobic influences into Scottish society. Nevertheless, the Parliament has shown itself able to separate church from state affairs despite enormous external pressure. As Iain Macwhirter has indicated, the Conservative Party anti-repeal motion in the Scottish Parliament was 'not about the promotion of homosexuality but about equality before the law'.[30] In relation to the Northern Ireland Peace Process, Tony Blair said, 'Part of this whole process is that we start to treat each other like normal human beings in a normal society.'[31] The Scottish Parliament is attempting to set a precedent for allowing for *difference* in society that perhaps Northern Ireland could take lessons from.

Editor's Note:
Since this chapter was written by Gerard Murray, devolution will be re-established in Northern Ireland at Midnight on Monday May 29 2000, following a close-run vote by the Ulster Unionist Council on May 27 (53% supported David Trimble's position).

Notes

[1]   BBC *On the Record*, 27 June 1999.
[2]   *The Guardian*, 23 May 2000.
[3]   Interview with Tom Hennessey, 3 April 1998 senior Ulster Unionist Party negotiator at the inter-party talks leading to Good Friday agreement.

4   *Scottish Affairs*, 26, 1999, The 1998 British-Irish Agreement: Power-Sharing Plus, 25.

5   1992 Series - Paper no 6, Scotland's Government - the transition to Independence, August 1996.

6   Gerry Hassan, 'Scotland's Parliament: lessons for Northern Ireland', *Democratic Dialogue*, 1998.

7   Clare O'Halloran, *Partition and the Limits of Irish Nationalism: An Ideology Under Stress*, Gill and Macmillan, Dublin, 1987.

8   Page 7, Paul Routledge, *John Hume A Biography*, HarperCollins, London, 1997. See also: Gerard Murray, *John Hume and the SDLP: Impact and Survival in Northern Ireland*, Irish Academic Press, Dublin, 1998.

9   Official Report of the Scottish Parliament, vol 4, no 7, col 629, 2 February 2000.

10   Official Report of the Scottish Parliament, vol 4, no 7, col 630, 2 February 2000.

11   Interview with Shona Robison, SNP, 8 September 1999.

12   Official Report of the Scottish Parliament, vol 4, no 7, col 638, 2 February 2000.

13   Paisley et al., 1982, 50 - 2. Quoted in John Whyte, *Interpreting Northern Ireland*, Clarendon, Belfast, 1991.

14   *The Irish Times*, 17 December 1999.

15   The Lothian Lecture, by the Taoiseach, Bertie Ahern, 29 October 1998.

16   See comments by Alex Salmond in *The Herald*, 24 August 1996.

17   Alex Salmond, 'Northern Ireland and the Scottish Question' *Scottish Affairs*, 13, 1995, pp. 68 -81.

18   Interview with Dan Mulhall, Irish Consulate, 31 August 1999.

19   *Belfast Telegraph*, 6 October 1999.

20   Figures supplied by the Statistics and Research Branch of the Department of Education in Northern Ireland.

21   Figures supplied by Higher Education Statistics Agency.

22   These areas where identitfied by the Taoiseach, Bertie Ahern at the inaugural meeting of the BIC on the 17 December 1999.

23   *Official Report of the Scottish Parliament*, vol 4, no 7, col 648, 2 February 2000.

24   *The Scotsman*, 23 December 1999, 'COSLA initiative for Irish council link-up'.

25   *The Guardian*, 10 August 1999.

26   *The Herald*, 10 August 1999.

27   Ibid.

28   *Scotland On Sunday*, 12 December 1999, 'Catholic paper in shame list of MSPs'.

29   *Evening News*, 16 December 1999.

30   *Sunday Herald*, 13 February 2000.

31   *The Independent*, 18 December 1999.

10 Scotland and the EU: All Bark and No Bite?

ALEX WRIGHT

Introduction

Scotland's relationship with the EU has long been a political minefield. Although it would be tempting to attribute this to success of the SNP's 'Independence in Europe' campaign, that is not the whole picture. Rather, the SNP and other critics of the previous constitutional arrangement merely exposed structural and procedural deficiencies that had existed for some time but that were all the more apparent as result of an intensification of European integration in the aftermath of the Single European Act (SEA) and the Treaty on European Union (TEU). Despite the principle of Subsidiarity, potentially Scots had insufficient influence over EU policy because decision-making resided within a European political arena from which political institutions in Scotland were for the most part at least one step removed [1]. As far as the (then) Scottish Office was concerned, the elongated chain of communication stretching between Brussels and Edinburgh had two significant ramifications. First, usually any input from the Scottish Office would have to dove-tail into a pan-UK position with the result that Scottish needs might be diluted or displaced altogether. Second, there was always the threat that the Scottish Office might not be consulted in time because of bureaucratic delays in London (or for that matter in Brussels) and there have been occasions when it was not consulted at all. Moreover, when officials from the Scottish Office attended inter-departmental meetings with the lead department to agree a position on an EU policy proposal, there was the likelihood that Scottish Officials could be out-ranked and over-ruled by colleagues in London[2]. In effect there was an absence of political autonomy for Scotland but of greater concern, however, was the deficit of ministerial leadership.

Even though the Secretary of State was a member of the Cabinet committee on EU affairs which dealt with strategic issues such as Enlargement and EMU there was no EU policy specifically for Scotland *per se*. In the absence of a European agenda Scotland's approach to the EU and its policies was piecemeal. Worse still, by adopting a centrist approach to government, successive Conservative administrations under Margaret

137

Thatcher and John Major failed to recognise that Scotland's priorities in the EU could be quite distinct at times from the rest of the UK. In turn that was undermined further by growing incompatibility between Scotland's needs and ministerial conduct in London. By adopting an increasingly Euro-sceptic stance during the 1980s and on into the next decade, the UK government was marginalised in the EU which in turn had repercussions for Scotland. In sum Scottish needs were no longer safe in the Government's hands but by virtue of the constitutional arrangement that pre-dated 1999 the dearth of democratic controls ensured that Scots lacked the means to hold Ministers to account for their lack-lustre performance.

Consequently Scottish Institutions and organisations were tempted to by-pass London and deal with Brussels direct. Although this strategy met with some success their influence was limited. This was because for the most part the member states remained the most potent actors, despite the formation of the Committee of the Regions and the growing use of co-decision by the European Parliament. The member states took the key decisions in the Council of Ministers and their administrations were usually responsible for overseeing the implementation of EU policies within their territories. The degree of territorial autonomy was dependent on the internal constitutional arrangements within each member state but commentators have rightly suggested that for the most part direct approaches to Brussels by territorial interests *complemented* rather than replaced governmental channels[3].

Over the years UK governments have attempted to improve the linkages between Scotland and the EU. After the SEA, the ministries in Whitehall were increasingly stretched, with the result that a host of EU related functions were decentralised to Scotland[4]. By the early 1990s the Conservative administration made increasingly desperate attempts to deflect the SNP. Ministers talked up the chimera of a 'multi-track' approach to Scotland's relations with the EU based on a mix of formal and informal linkages (including the formation of Scotland Europa in Brussels). These developments were archetypal of the incremental response by the Government of the day to Scottish demands for greater autonomy. But by 1996 the Government had failed to disguise the fact that in terms of outcome Scotland fared little better than other parts of the UK or some might argue, it fared worse. During the 1990s the BSE crisis was mishandled, there are doubts about how much EU aid was actually allocated to Scotland, the fishing industry failed to secure sufficient funding either for decommissioning or for modernising the fleet. Edinburgh was prevented from fighting its corner for an off-shoot of the European Central Bank and the Whisky industry had to endure successive increases in duty in the UK, whilst campaigning for a level playing field in the EU. All-in-all by the mid-1990s it was difficult to avoid the impression that Scotland's representation in the EU was inadequate[5].

Now that Scotland has its own parliament Scots are tempted to believe that Devolution will result in more influence over relations with the EU[6] but that may prove optimistic not least because this is 'reserved' to Westminster. That said, the EU was assigned its own chapter in the White Paper and in so doing the Government acknowledged that despite its reserved status where possible the Scottish Executive and Parliament would have an 'important role' concerning EU affairs[7]. If the White Paper portended much, the reverse was so with *The Scotland Act* itself, where EU relations warranted the briefest of mentions[8]. Conversely when the Consultative Steering Group (a body whose primary remit was to devise the parliament's procedures and standing orders) issued its proposals, it was evident that its authors expected that the Scottish Parliament would be proactive in relation to EU policies[9]. Just over a year later when the Memorandum of Understanding (MoU) and the Concordats were published, setting out the relationship between Holyrood and Whitehall, the pendulum swung the other way - they appear to constrain the potential for Scottish autonomy as we shall see below.

Yet the 'Joint Ministerial Committee' (JMC) - the details of which were published in tandem with the MoU - is potentially significant. It will act as a forum whereby Scottish ministers will discuss EU-related issues with colleagues from Whitehall on a more equitable footing than was the case with their former colleagues at the Scottish Office who acted as 'bargain hunters' within the UK political arena. Even so, if the two levels of government were to disagree over EU policy, the final say rests not with Edinburgh but with London. Given the track record of UK Governments, it would be understandable if sceptics viewed the new arrangements for EU representation as somewhat ephemeral and as such it is merely a continuum of the incremental approach that pre-dated Tony Blair's premiership.

A Continuum of Incrementalism

At first glance the new Executive is little better off, as regards representation in the EU than its forebear. Ministers from the Executive may attend the Council of Ministers but there is nothing new in that as ministers at the Scottish Office have been present where a particular issue or area of policy has ramifications for Scotland (e.g. fisheries). Yet according to the Concordat they do not have the automatic right to be there, they can only be present with the assent of their ministerial counterparts from the lead department. As before should they vote in the Council they represent the UK at that moment in the sense that they vote for the UK in its entirety, not for Scotland in particular. Furthermore informed opinion suggests that in practice the likelihood of Scottish ministers voting will be quite rare. The situation for civil servants is less clear-cut as we shall see, but they can still be out-ranked by colleagues in London and for the most part the Scottish view will still have to be imbued within the overall UK position on EU policy.

As far as MSPs are concerned the Parliament's European Committee will ensure that the Executive's handling of EU issues will not be neglected but its influence over policy is questionable. Under the CSG's proposals Holyrood's committees were supposed to have much greater influence over the formulation of policy than is the case with the committees at Westminster. But that does not apply in this instance as competence for policy resides at the European level. Moreover, both the committee and its counterparts at Westminster suffer from the same basic flaw. These bodies are supposed to scrutinise EU policies before they come on the Council of Ministers' agenda and should they raise an objection to a proposed policy then this would be conveyed to whichever minister was representing the UK in the Council. All too often however, a deal has been cut in the Council well before it reaches the parliamentary committees and although there have been attempts to improve this deficiency, by and large the parliament at Westminster has been largely reactive towards developments in the EU. There is little to prevent Holyrood suffering the same fate and this is exacerbated further by the reserved status of the EU. At the time of writing there is no evidence that the Scottish committee's 'Opinions' will be transmitted direct to Brussels – as its members were surprised to discover in August 1999. Instead one or a combination of channels will be at the committee's disposal. These include; its counterpart in the Commons (in readiness for the formulation of a UK view), the Scottish Executive (for re-transmission to Whitehall) and - if such an office were to remain *in situ* - the Secretary of State (so that an Opinion can be relayed to the Cabinet and/or the PM) [10]. So, there is nothing to prevent the committee's views being diluted long before they reach the EU and even then they may arrive too late in the day because the Council has already reached agreement.

Scotland House in Brussels attracted widespread publicity when it was established in the autumn of 1999. This is not a Scottish Embassy as such but it does mark something of a sea-change as Scottish civil servants will now have a permanent base in Brussels. Moreover they will probably be co-located with the residents of Scotland Europa with the result that staff from both organisations will have the opportunity to share intelligence and pool their resources. That said the question remains as to how much difference Scotland House will really make. First Scotland Europa has been up and running since the early 1990s and over that time it has earned a reputation for keeping its ear to the ground and promoting Scottish interests. Will six civil servants under the same roof transform things? The number of officials is markedly smaller than the equivalent bureau for Bavaria for example. Second, it could be argued that permanently based staff in Brussels may not amount to much anyway because by the early 1990s on average there was a Scottish civil servant in Brussels every working day of the week and that does not include reciprocal visits to Edinburgh by EU officials. Likewise, for many years Scottish civil servants have been seconded to UKRep, the UK's permanent diplomatic mission in Brussels, and where necessary they have attended working groups in the Council of Ministers[11]. It could therefore be inferred that once again little has changed in substance.

At this point in time when the Parliament is in its infancy, there are grounds for claiming that Scotland's representation in the EU remains inadequate. The UK executive retains too much discretion over how Scottish affairs should be promoted and protected in Brussels and Scotland remains one step removed from the Council of Ministers which in spite of co-decision with the European Parliament and institutional reform continues as the EU's power house. But now that there really is a Scottish Parliament could Scotland's autonomy in the EU be transformed in all sorts of ways which had not been foreseen by the architects of devolution in the Labour Government?

Pressure for Change Post 1999

For a variety of reasons the situation for Scottish ministers and civil servants will be profoundly different than has been the case hitherto. In former years ministers could blame colleagues in London and the Secretary of State could use Cabinet consensus as a smoke-screen (albeit that the post holder could wield considerable influence behind the scenes on behalf of Scottish interests). What is so different now is that having campaigned for a parliament and with public expectations so high about Scotland's influence (including over EU policies), Ministers cannot afford to be seen as powerless. Given its high profile, whether the Executive likes it or not, relations with the EU will be one of a number of litmus tests as to whether constitutional reform amounts to territorial empowerment. Regardless of its status as a reserved power, Scottish ministers will need to be seen to be more effective when fighting Scotland's corner than was the case before 1999.

As ministers begin to face the flak Scottish civil servants will feel the fall-out. Yet in theory their autonomy from London is circumscribed by the Scotland Act and the same applies somewhat less formally to the Concordats, the JMC and its associated 'action committees (these arrangements are not legally binding and they are subject to review when necessary). An underlying objective of the Concordats is to ensure that there is a degree of coordination between civil servants in Edinburgh and London regardless of which parties are in Office north or south of the Border. Consequently both wings of the civil service will be given advance notice of policy developments in order to prevent one side or the other from being wrong footed. But could their impartiality or for that matter their desire for cohesion withstand the *real politic* of the Scottish arena post 1999? In the short term it could be supposed that there is little room for conflict between a Scottish Executive and the UK government. Labour is in Office and its leaders have an alliance of sorts with the Liberal Democrats at the UK level and there is a more formal arrangement at the Scottish tier where the Lib-dems enjoy a measure of autonomy from their colleagues in London. So far so good but there have already been rumblings of discontent amongst the Liberal Democrats over fishing zones and this was amplified in December 1999 when Gordon Brown announced the new Ministerial Joint Action Committees without apparently consulting the Lib-dems at all. That though would be no more than a warm up for the dislocation, if the SNP eventually became the Majority Party in Holyrood.

Some, especially those south of the Border, have been tempted to believe that the avoidance of such a political fracture was one of the primary functions of the MoU and Concordats. At the time of writing that appears to be untrue. One of the reasons why the Concordats and MoU were to be subject to review was that it would be necessary to take into account political developments in both Scotland and the rest of the UK as time went by. If the SNP attained a majority in Scotland on a mandate to opt for independence, the wishes of the electorate would take precedence over the Concordats and MoU. In the meantime a problem could arise, however, if one day the SNP were to form a coalition government with say the Lib-dems and its leader held the portfolio for the EU. If by chance that individual was called upon as minister for EU affairs in Scotland would he or she be prepared to represent the UK as well, given that their status as a nationalist politician? To-date the Scottish Executive do not have an EU minister, as this portfolio has been parcelled out across the Executive. In so doing it raises the question as to whether the current political leadership in the Executive wants to down-play any notion of Scottish autonomy in the EU. But how long will this be tolerated by MSPs and more particularly by the electorate in Scotland?

Nor should Scotland House be discounted. High-grade intelligence can be invaluable in all sorts of ways. Conceivably the most clear cut is that Scottish officials can lobby EU officials well before legislation is even drafted and for its part the Commission could consult Scotland House informally to assess what the Scottish position might be with regard to a proposal on environmental policy for example. Although Scottish Officials have been seconded to UKRep in the past, formally they become members of the FCO not Scottish Officials. Now there will be a dedicated desk for the Scottish executive at UKRep which will enable its personnel to access UKRep's data-base when necessary. By housing civil servants and the residents of Scotland Europa in one building there will be cross fertilisation of ideas and intelligence. More particularly as was intimated by one senior civil servant in his testimony to the European Committee on August 18 not only will Scotland House become a 'focal point' for Scottish interests in Brussels, it will be multi-dimensional. He commented,

> Scotland House is intended to be a focus for Scottish interests. From the Executive's point of view, we are riding two horses. We want to be a Government office and to be regarded as something official, but, equally, we want to be part of the larger whole. I suspect - I am being slightly speculative - that over time Scotland House will appear in different guises depending on who is looking and with whom they are talking.[12]

For their part the MSPs and the European Committee should not be under-estimated. Whilst it has been claimed above that it will be reactive to developments in the EU it can still be extremely potent. First it can call Scottish ministers to explain their position on a particular EU policy and it can examine their conduct in considerably more detail than would be the case on the floor of the House. In addition the committee can act as a trip wire by identifying potentially contentious developments ahead, such as the allocation of EU monies to the UK and their re-allocation across its constituent territories - as was the case at its 3rd meeting on August 31 1999 when J.McConnell the Finance Minister answered questions on the structural Funds[13]. Even though it is a reserved power there is nothing to prevent MSPs from holding a debate on Scotland's relations with the EU - as was so on November 10 when the First Minister moved 'That the Parliament endorses the Scottish Executive's policy of continued positive engagement within the European Union...' Although the motion was passed this did not prevent Alex Salmond, the SNP leader, from asking why there was no Scottish Minister with a portfolio for the EU[14]. Furthermore, if they so chose, MSPs could pass a resolution condemning HMG's handling of EU affairs; had the parliament been in existence in 1996 that could well have occurred as the BSE crisis unfolded. So in a narrow sense the committee could be a source of pressure and when allied with MSPs more generally it might induce ministers to adopt a more strategic approach to Scotland's agenda in the EU.

Scotland's Agenda in the EU

The difficulty for policy makers is deciding what exactly should comprise Scotland's priorities in the EU. Should they opt for a continuum of the previous multi-track arrangement, whereby there was relatively little co-ordination by government and Scottish actors were left to plough there own furrow? In a similar vein the Scottish Executive could opt for the 'shotgun' approach. In effect it would set out to influence any EU policy proposals or developments that had ramifications for Scotland. The difficulty with that is not only would it stretch the Executive's resources, it would also run the risk of being reactive rather than pro-active. Alternately Scotland's political leadership could identify specific interests/policy fields that are of particular concern north of the Border and it could 'cherry pick' those with the highest profile. Economic development is one that immediately comes to mind; so securing EU aid might be treated as a priority. The Fishing and Agriculture sectors are of considerable importance to rural and coastal communities away from the central belt. For their part the crofters have no comparable group elsewhere in the UK; they are distinctively Scottish. Although Edinburgh's status as a financial centre falls well short of the city of London, it is a substantive player in the European context. The whisky industry is a major export earner not just for Scotland and by default the UK but also for the EU. Furthermore it too is a valuable source of employment in some of the more sparsely populated areas of Scotland. Then there is the energy sector. With the oil sector set to decline the technological expertise that has accumulated in the North and the East of Scotland should be put to good use elsewhere in the EU and beyond. More latently the Executive might concentrate on building a sound relationship with the EU by stressing its commitment to European integration (as was the case with the Motion it put before Parliament on November 10).

Bearing in mind that for the moment the Executive's political leadership comprises a coalition this could contribute to policy incoherence regarding the EU. For instance supposing Labour eventually postponed the holding of a referendum on UK entry to EMU for 5 years or more would that be acceptable to the Liberal Democrats? Equally over the years each of these parties has adopted radically differing positions on European integration. The Liberal Democrats have been the most constant in so far as they have favoured a Federal EU. Labour has blown hot and cold; during the 1980s the party campaigned for the UK to leave the EU but by the early 1990s it was much more supportive. Even so, there is mounting scepticism in Labour's rank and file over the EU, and there is no sign that the Government would support deeper integration if it led to harmonisation of taxation. If both parties were to differ substantially over the future direction of the EU it would be difficult to see how that would not filter downwards to their Scottish wings. Similarly will the Executive and the Parliament have a singe view on what exactly should constitute Scotland's European agenda? It will be interesting to see. To date the Conservatives have been the most hostile towards the EU, whilst the SNP maintains that the EU should be a confederation of states rather than a federation, which itself may be problematic given the commitment to 'Ever Closer Union' in successive European treaties. Will the Executive and Parliamentarians be tempted to pursue short-term spin-offs in a bid to curry favour with the media and the electorate? Will they opt for 'grant getting' more than setting out to influence the EU's agenda? Will they have the political savvy to play the long game and contribute further to Scotland's reputation as a mainstream player in the EU? The answer to such questions will depend on a number of factors but one of the most crucial will be political leadership. Will it be weak or strong or given the circumstances will Labour's leadership in London *allow* it to be strong as regards EU affairs? When the author asked the First Minister what Scotland could bring to the EU he replied that he would not be so conceited as to consider such a scenario[15]. An adroit answer from a deft politician perhaps, but it cultivates the impression that under new Labour, Scotland must not be seen to be pursuing a partisan agenda in the EU.

To the observer the overall impression is the just how fluid things have become and the extent to which in the absence of a written constitution we have entered uncharted waters. This is particularly apparent concerning the office of Secretary of State. There are a number of possible outcomes; the post might continue, or eventually a single Cabinet minister could be responsible for all the territorial governments in the UK, or the respective Secretaries of States might be significant players in the newly formed 'Joint Ministerial Committee'. What is particularly notable, however, is the low profile the present incumbent, has assigned to EU affairs. At the time of writing there have been no speeches from Dr John Reid on the subject and no publications have been issued by the Scotland Office on the EU[16]. Yet surely as it is a reserved power then it would follow that the Secretary of State would have some involvement in the EU. A generous conclusion would be that he has no desire to usurp the First Minister after they apparently fell out during the summer of 1999. A more likely reason is that Scotland's relationship with the EU is far too thorny a subject for Dr Reid to tackle given the ambiguities of competence and it is therefore best avoided altogether. However the silence of the Scotland Office is cause for concern because it suggests that here too no EU strategy is being developed on behalf of Scotland.

Questions for the Future

One imponderable is whether or not the EU will assign more power to its regions and stateless nations. Despite the principle of Subsidiarity at present the relationship between the two is asymmetrical by virtue of the fact that is it primarily up to the member states to decide individually whether or not their territorial governments can become more directly involved in decision-making in the Council of Ministers. Even then, those territorial governments that have voted in the Council are *de facto* representing their member states. There is little evidence that there will be further empowerment of the EU's territorial governments at the next IGC - this will have ramifications for the future of Scottish politics. If the EU does not devolve more power to its 'regions and stateless nations, and if it remains the preserve of the governments of the member states and if more power is assigned to the EU over the next few years with the result that considerably more laws are created at the European rather than at the UK level, there will be growing pressure for Scotland become a member state in its own right. In effect relations with the EU is the Achilles heel of devolution.

Before a Scottish Government takes the first steps towards independence, it may attempt to join or construct alliances with other devolved bodies in the EU. Trans-national networking has long been a feature of EU politics and the EU's territorial governments have been particularly involved in such alliances because this has enabled them to lobby more powerfully for EU funding. Certainly the creation of the Scottish Parliament has aroused interest on the Continent. In the first few months of its inception, delegations have visited from the Länder and some territorial heads of government have openly suggested that the Parliament might re-invigorate the push for the 'regionalisation' of the EU. Certainly there is nothing to prevent the Parliament from linking up with other devolved bodies in the EU - both the White Paper and the Concordats view such a development as potentially worthwhile. There are signs that the EU's territorial parliaments will establish their own umbrella organisation and it possible that one day this might usurp the EU's Committee of the Regions.

Yet, within the Executive there is the concern that if too much effort was expended on building links with third countries and/or territorial governments in the EU this would take up valuable resources when civil servants are already fully stretched (e.g. dealing with more and more Parliamentary Questions and a raft of new legislation). Similarly doubts have been aired over whether MSPs would really have the time to indulge in the creation of a 'Europe of the Regions', when the electorate is more concerned with domestic issues such as transport, health and education. So although there is some support for functional alliances that may produce extra funding or a trade-off on fish quotas, there is less enthusiasm for overtly political territorial networks that yield little in return in the short term.

Then there is the issue of inter-governmental relations within the UK and beyond. Now the North of Ireland has its own devolved government there is no reason why it and its counterpart in Wales could not form a caucus with Scotland on EU policies - for example on the environment or agriculture. If so they might form a powerful alliance at the territorial level within the UK. Even though neither party is yet in government both Plaid Cymru and the SNP have jointly expressed their concerns to the European Commission over the UK government's attitude to Additionality concerning the allocation of EU funds - an issue that helped bring down Alun Michael, the Welsh First Minister during February 2000. As Gerard Murray suggests in the previous chapter, theoretically there is nothing to prevent a Scottish Government aligning itself with the South of Ireland over a particular issue. That might apply such as the 2002 review of the Common Fisheries Policy, for example.

The Challenge of Devolution

Superficially little appears to have changed with regard to Scotland's political relationship with the EU. It is still one step removed from decision making in the Council of Ministers, thanks to the Scotland Act and the Concordats that followed. As was the case before the parliament came into existence, there is too much reliance on informal mechanisms; there needs to be considerable collaboration between the UK and Scottish tiers of government if the outcome is to be mutually satisfactory. In particular it will be especially dependent on there being a UK Government in office that is sensitive to Scottish needs in the EU, that is willing to respond to those needs, that possesses the capacity to influence developments at the European level. Whilst this may be so to some extent at the time of writing, constitutionally there is little to prevent a re-run of the 1980s and 1990s when the reverse applied under the Conservative Governments of Margaret Thatcher and John Major.

Nonetheless, even though European affairs is a reserved power the existence of a Scottish Parliament promises much in the future. Ministers will be kept on their toes and they will not be able to 'hide behind London'. They will be encouraged to be pro-active towards the EU and their actions will be subject to the scrutiny of MSPs both in the European Committee and in the debating chamber. The Parliament has the potential to undermine the integrity of the UK in as much MSPs might adopt different position on developments in the EU to their colleagues at Westminster. Accordingly we can expect the UK Government to adopt a more consensual style with regard to its territorial governments if it wished to avoid the impression amongst the other member states in the EU that it was divided internally.

Yet, the question remains as to whether the existing constitutional arrangement is little more than a halfway house. It is already evident that there is a head of pressure from MSPs for greater influence over EU policy than has been the case up until now. It is questionable whether this can be contained under a devolved system of government where relations with the EU remains formally a reserved power, but with a multiplicity of informal mechanisms that are designed to accommodate Scottish demands within a pan-UK position. With such a high proportion of legislation now emanating from the EU, some of which is especially contentious for Scotland (e.g. the CAP and the CFP), it will be near impossible for the UK and Scottish governments *not* to have substantive disagreements at some point in the future. MSPs and ministers can 'bark' as much as they like but if they have too little 'bite' then devolution may well be no more than a staging post for an 'independent' Scotland in the EU.

Notes

1   Scott et al (1994), Subsidiarity: A 'Europe of the Regions v the British Constitution', in *The Journal of Common Market Studies*, vol 32 no. 1 March 1994, pp. 47-67.

2   Wright, A. (1998), PhD, *Scotland and the EU: a case of Subsidiarity or Dependency?*, University of Dundee.

3   Hooghe, L. (1995), Subnational Mobilisation in the European Union, in *West European Politics*, Vol 18, July 1995 No 3, pp 175 – 198.

4   Wright, A (1995), 'The Scottish Office and the EU', in S. Hardy, M. Hebbert and B. Malbon in *Region Building*, the Regional Studies Association, London.

5   Wright, A, (1998), PhD *Scotland and the EU: a case of Subsidiarity or Dependency?*, University of Dundee.

6   Brown, A. et al (1999), *The Scottish Electorate*, Macmillan, London, 141.

7   HMSO (1997), Cm3658, Scotland's Parliament, 5.1.

8   *Scotland Act* 1998, Chapter 46, Schedule 5, 7 (1).

9   'Furthermore, the Committee should be prepared to take a proactive role in the development of key areas of EU policy' (Scottish Office 1998, The Consultative Steering Group of the Scottish Parliament, Section 3.4. 30.3).

10   Scottish Parliament, European Committee Official Report, Vol 1 No 2, August 18 1999 Cols 31-2.

11   Wright, A. (1998), *Scotland and the EU: a case of Subsidiarity or Dependency?*, University of Dundee.

12   Scottish Parliament, European Committee Official Report, Vol 1 No 2, 18 August 1999 Col 46.

13   Scottish Parliament European Committee, Official Report Meeting 3, 31 August 1999, Col 88.

14   MINUTES OF PROCEEDINGS Vol. 1, No. 30 Session 1, Meeting of the Scottish Parliament, Wednesday 10 November 1999.

15   Questions invited from the audience after a speech by the First Minister at the Annual Conference of the European Movement, Perth, Saturday, November 6 1999.

16   Information based on the Secretary of State's Website www.scottishsecretary.gov.uk

11 An Oxymoron: The Scottish Parliament and Foreign Relations?

TREVOR SALMON

It Ought to be Clear But ...

The Scotland Act 1998 appears to answer the question of the role of the Scottish Parliament and Executive in foreign policy, foreign, international, or external relations very clearly and directly. Schedule 5 .7.- (1) entitled 'Foreign affairs etc.' says

> International relations, including relations with territories outside the United Kingdom, the European Communities (and their institutions) and other international organisations, regulation of international trade, and international development assistance and co-operation are reserved matters.

Elsewhere matters identified by Schedule 5 as 'reserved' include
'The defence of the realm'
'The naval, military or air forces of the Crown ...'
'Treason...'
'Fiscal, economic and monetary policy, including the issue and circulation of money, taxes, and excise duties... the exchange rate and the Bank of England'
'The currency ...'
'Money laundering ...'
'Immigration and nationality ... including asylum ... free movement of persons within the European Economic Area; issue of travel documents.'
'National security, interception of communications, official secrets and terrorism ...'
'Extradition.'
'Competition. Regulation of anti-competitive practices and agreements, abuse of dominant position; monopolies and mergers'
'Control of nuclear, biological and chemical weapons and other weapons of mass destruction.'
'Regulation of activities in outer space.'

although

> financial assistance to commercial activities for the purpose of promoting
> or sustaining economic development or employment

is *not* reserved.

Elsewhere in the Act it is made clear that the Parliament and
Executive do have to comply with and fulfil the obligations of European
Communities membership and the European Convention of Human Rights.

Thus, the role of the Scottish Parliament and Executive in
international relations appears to be clear: it is nil. It is a reserved area,
where all the 'reserved matters' are carefully enumerated. This, however, is
clearly an oxymoron, since it is already outmoded both empirically and
theoretically. Given the nature of the contemporary political system, even
the provisions of the Scotland Act *cannot* rule out the Scottish Parliament's
involvement in international or foreign relations.

The Context of War

It is sublimely ironic that as Iain Macwhirter wrote in April 1999

> WELL, what no-one expected was a khaki election. The campaign to elect
> Scotland's first parliament in 300 years has inevitably been eclipsed by
> the conflict over Kosovo - and not just because of Alex Salmond's
> condemnation of Nato airstrikes. War - even undeclared war - changes
> everything. The common currency of party political debate becomes
> devalued when matters of life and death are at stake (Sunday Herald 11
> April 1999).

While the force of outside events could not be excluded from the campaign,
they doubly intruded given that Alex Salmond, the leader of the Scottish
National Party, chose to condemn British and Nato policy at the outset of
the election campaign, when he was given the right to reply to a Prime
Ministerial broadcast. Salmond accused Prime Minister Blair and NATO of
pursuing a 'misguided' policy of 'dubious legality and unpardonable folly'.
He went on to apparently compare NATO bombing to the Nazi blitz on
London and Clydebank in World War Two (The Scotsman 30 March
1999).

Whatever the specific merits of the argument, the broadcast had
important consequences. It led to at least the first half of the election
campaign focusing on his words and the response from Belgrade (which

used it as a propaganda tool); and it clearly lost him support, not just because a third of his own party supported the air strikes but because about a quarter of the electorate claimed it would make them less inclined to vote SNP. This in itself was important given that the final outcome produced no single party overall majority. Muslim opposition to his stance may even have contributed to the SNP failing to gain a key Glasgow seat they had targeted. In addition, it raised questions over Salmond's judgement and statesmanship.

The khaki election and the 'unpardonable folly' remarks ensured that the classic components of international relations - war and the use of force and power - would have a profound effect on the Scottish body politic not just in the election itself but in the life of the first Scottish Parliament the electorate voted for at the end of the campaign.

These episodes demonstrate the difficulty of maintaining the alleged separation of international relations from mainstream politics. Whatever the Scotland Act says, it is difficult to see how the Parliament can avoid at least debating such major issues in the future. As early as May 1999 Alex Salmond himself called for the Parliament to debate the consequences of NATO and British policy, despite the fact that it had no responsibility for foreign affairs. At that time, however, the Parliament was not fully functioning and MSPs were not allowed to put down motions, so the issue was moot. Such issues are unlikely to be moot in the future. Indeed, when the Scottish Affairs Committee of the House of Commons investigated 'The Operation of Multi-layered Democracy' in 1998, Anne Begg MP asked Dr. Charles Jeffrey, an expert witness:

Do the sub-national assemblies debate reserved powers ...?

Dr. Jeffrey:

Yes, they do debate reserved powers. They can debate anything they like (The Operation of Multi-layer Democracy, Scottish Affairs Committee Minutes of Evidence February 1998 col.51).

When finally published in the autumn of 1999 the Memorandum of Understanding between Her Majesty's Government and Scottish Ministers and the Welsh cabinet accepted that:

The devolved legislatures will be entitled to debate non-devolved matters, but the devolved executives will encourage each devolved legislature to bear in mind the responsibility of the UK Parliament in these matters (Scottish Executive Memorandum of Understanding and supplementary agreements).

The Anachronistic Nature of the Separation the Scotland Act purports to Uphold

The foundation of international relations lies in the fact that there are many different societies or groups of individuals scattered all over the globe, and these scattered societies have some form of contact with one another. International relations, even foreign relations, involve the study of the interactions that take place between these societies or entities, and the factors that effect those interactions. This broad definition allows for the time before the emergence of the modern state system, and also for the potential situation where the state is superseded by some new form of government. For much of the twentieth century, international relations has actually been concerned with the study of relations between states. However, this has come increasingly into question given the emergence of entities such as the European Union, which is neither a state nor a nation, and of 'sub-national actors' (SNAs) or as Hocking prefers 'non-central governments' (NCGs) (Hocking 1993). Both these developments plus technological change, global market forces and changing social patterns, have affected the nature of the actors in international relations at all levels.

The very word 'international relations' infers of course a particular concern with relations between nations, but it does not have to be so confined. For most of the twentieth century, the main actors in international relations were seen as states, but 'relations' implies all the various types of interaction between geographically separate entities, whether these be governmental or private. They range from the great issues of war and peace to tourism, trade, commerce, communications and even letters that cross state boundaries. Thus there is no particular characteristic specified as to the nature of the interaction. Thus contrary to the narrow traditionalist realist view of international relations and foreign policy/relations, which focuses on the physical security and protection of the territory of the state and its people, one needs to look wider.

The traditional view implicit in much of international relations literature follows the assumption of Bodin that social and political order and legality are the highest values of a society and that it followed, along lines pursued by Hobbes, that in every given territory sovereignty must be united in one clear, secular authority to establish and maintain order. Following this, the Treaty of Westphalia and the evolution of international diplomacy since 1648, there was a situation where sovereign rulers exercised theoretically and actually the unchangeable right of law-making. Law-making and foreign relations were for centuries the preserve of the state. Thus only those holding authority on behalf of states could have political relations with each other. A generation ago, as Wallace noted, it was still possible for some to assume that:

> The separation of foreign policy from domestic policy is fundamental to the traditional concept of the nation-state (Wallace 1971 p 8).

Or as Henry Kissinger put it:

> In the traditional conception ... the domestic structure is taken as given; foreign policy begins where domestic policy ends (Kissinger 1969 p 261).

Foreign policy was seen as providing the bridge between domestic and international politics. It was seen as state-centric. But in the intervening period the increasingly indistinguishable separation between domestic and international politics has been widely, if not universally accepted. Wallace himself and others were coming to the view that

> The difference between 'national' and 'international' now exists only in the minds of those who use the words (Philip E. Mosely 1961 pp 43 ff.).

not least because as Rosenau observed:

> In certain respects national political systems now permeate, as well as depend on, each other and that their functioning now embraces actors who are not formally members of the system. These non-members not only exert influence upon national systems but actually participate in the process through which such systems allocate values... Most important, the participation of non-members of the society in value-allocation and goal-attainment is accepted by both its officialdom and its citizenry (Rosenau 1980 p 146).

These are 'penetrated' political systems, that is, a political system in which:

> Nonmembers of a national society participate directly and authoritatively, through actions taken jointly with the society's members, in either the allocation of its values or the mobilisation of support on behalf of its goals (ibid pp. 147-148).

Increasingly it is apparent that the number of government departments (central, regional and local) which may wish to conduct relatively low-level negotiations or to exchange technical information, with their opposite numbers in other states is virtually equivalent to the number of departments which actually exist at any given time. It has certainly become increasing true that each major department of central government has its 'foreign ministry'. As Roy Jones noted a consequence is that

> They develop their own norms and procedures, their own styles of cross-national conduct ...The extension of this process has the effect of gradually reducing the role of politics in world society... the role of diplomacy is gradually supplanted by the activities of sets of non-technical experts.... (Jones 1970 p 147).

It is thus not surprising that terms such as micro, para, proto, and pluri-national diplomacy have entered the conceptual language in the attempt to categorise the emerging international role of SNA's and NCG's, although it is clear that both the concepts and what they refer to are still contested (Aguirre 1999).

It is clear that the Weberian definition of a government, portraying it as a formal state structures granted with legitimate authority over a society inhabiting a defined territory, is increasingly being eroded, both by the involvement of sub-national as well as supranational actors (Rhodes 1997). The government described by Weber cannot function in the contemporary world, due to the magnitude of internal and external factors forcing it to be more flexible. More contemporarily, there is now recognition that there is:

> A complex, multi-layered, decision-making process stretching beneath the state as well as above it; instead of a consistent pattern of policy-making across policy areas, one finds extremely wide and persistent variations (Marks 1992 pp 221-223).

According to Marks multi-level governance explains systems of policy-making 'among nested governments at several territorial tiers', which take place as:

> The result of a broad process of institutional creation and decisional reallocation (Marks 1993 pp 391-410).

The capacity of SNAs or NCGs to participate effectively in the new multi-level governance system depends:

> Not merely on constitutional structures but on the nature of territorial civil societies economic resources, the capacity for political mobilisation and the capacity to project the territory internationally, especially within international regimes (Keating 1992 p45).

Clearly Scotland does have a significant resource base (financially and administratively) to allow it both to operate as and to become a more significant international actor, as long as one moves from the narrow view of international or foreign relations. Thus the simple propositions in the Scotland Act will not do.

The Scottish Parliament and Executive and International Relations

The British Government acting outwith the Scotland Act has itself acknowledged that there is an international role for the Parliament, namely its responsibilities and role in the 'British-Irish Council' that was part of the 'The Agreement' arrived at on Good Friday 1998 between the various parties involved in the attempt to resolve the Northern Ireland crisis. In the agreement and subsequently, Her Majesty's Government have accepted that the Council, which has met, comprises:

> Representatives of the British and Irish Governments, devolved institutions in Northern Ireland, Scotland and Wales, when established, and, if appropriate, elsewhere in the United Kingdom, together with representatives of the Isle of Man and the Channel Islands (The Agreement 1998).

And that the Council may meet at 'summit level', and in:

> Specific sectoral formats on a regular basis, with each side represented by the appropriate Minister; in an appropriate format to consider cross-sectoral matters.

In addition, it is to:

> Exchange information, discuss, consult and use best endeavours to reach agreement on co-operation on matters of mutual interest within the competence of the relevant Authorities.

The 1998 Agreement states that :

> Suitable issues for early discussion in the BIC could include transport links, agricultural issues, environmental issues, cultural issues, health issues, education issues and approaches to EU issues. Suitable arrangements to be made for practical co-operation on agreed policies.

Although:

Individual members may opt to participate or not in such common policies and common action,

It will be open to the BIC:

To agree common policies or common actions

And:

It will be open to two or more members to develop bilateral or multilateral arrangements between them. Such arrangements could include... mechanisms to enable consultation, co-operation and joint decision-making on matters of mutual interest; and mechanisms to implement any joint decisions they may reach. These arrangements will not require the prior approval of the BIC as a whole and will operate independently of it.

There was also the possibility of an interparliamentary links.

Although it might seem as if the BIC is a version of domestic politics, that interpretation is insulting to the Irish Republic, and it is instructive that the Irish Republic opened a Consulate-General in Edinburgh in September 1998; that Bertie Ahern, the Irish Taoiseach (Prime Minister), visited Scotland twice in 1999, in addition to visits by the Tanaiste (Deputy Prime Minister) and Foreign Minister. In November 1999 President McAleese visited Scotland. Just days earlier, the Irish government gave Donald Dewar, the First Minister, on the occasion of his visit to Dublin, the full trappings of a state visit (although technically it was not) with a six car cavalcade led by two police outriders. While this treatment was related to the delicate situation in the peace process and the Irish attempt to influence Unionist perceptions of Dublin, it does point up that the Scottish role in the BIC gives the new Executive and Parliament a very significant international responsibility and an added responsibility because of the ties between Scotland and Northern Ireland. In the first year of the new Scottish administration, other Scottish ministers also visited Dublin, including for example, Wendy Alexander, Minister for Communities, who was accompanied by deputy minister, two civil servants, one private secretary, a press officer and a visitors' officer from the British Dublin Embassy, as she sought to learn more about Dublin's anti-poverty strategy.

Edinburgh is now home to 13 full-time career consuls and to the representatives of about 40 different states. The most recent consulates to be established include the Irish, Taiwanese, Mongolian and Indian. The Indian Consul referred explicitly to the need to have:

Close contact with members of the Scottish parliament and executive.

While the honorary consul of Austria has referred to governments wanting to be:

As close as they can to the seat of power (The Scotsman 10 November 1999).

While some of the consulates have been in Scotland for some time, with some in Glasgow, the creation of the Holyrood Parliament has had an impact on both their numbers, their moving to Edinburgh and their modus operandi.

This raises the issue of whether in addition to 'Scotland House' in Brussels, there will need to be 'Scotland' Houses elsewhere. The Labour regimes in Edinburgh and London are cautious because of the association in some minds of overseas representation with diplomatic recognition and statehood. The Foreign Secretary, Robin Cook, has pointed out that the United Kingdom has 221 embassies around the world, at least four times the Irish number (Ireland being a model the SNP often likes to cite). He also noted that Ireland had no diplomatic representation in Hong Kong, Oslo, Sofia or Skopje, nor in states like Slovenia. Nonetheless, overtime the question of the representation of specifically Scottish interests is bound to arise, not least because Scotland has key economic interests to protect and advance, such as agriculture, tourism and energy. There may well be potential for friction with London here, since such offices would surely be tasked with specific responsibilities such as promoting tourism to Scotland and inward investment. Whatever the current arrangements, or the initial agreement in the Concordat on International Relations (see below), there is certain to be evolution in these areas, not least because MSPs will be answerable to a local electorate, and members of the Executive will be answerable to the MSPs. After a while, it is difficult to see how either can hide behind the provisions of a Concordat, especially when it comes to attracting jobs to or defending jobs in local areas. Even before devolution Ministers were asked in the Commons about when they would meet representatives of Scotland's financial services industry to discuss Edinburgh's position or the CBI in Scotland to discuss prospects for investment in Scotland.

So far some broad criteria for establishing Scottish links with 'foreign' actors have been identified, and they include:

• Shared interests in the economic, social, cultural and environmental fields;

• Shared geographic and demographic similarities (peripherality and population sparsity);
• Exchanges of information and expertise on the role of regional legislatures;
• Good potential to bring benefits to Scotland without prejudice to UK interests, including access to new markets or funding, achieving added insight into common issues and increased Scottish influence.

Only Marginal?

It is important not to over emphasise the significance of the external dimension. In the early life of the Scottish Parliament it has become clear that these international possibilities can be exaggerated. Although, as will be shown below, issues with an international dimension have been raised in the Chamber, there has actually been more evidence of a tendency noted with respect to the House of Commons and public opinion in the 1950's, when George Jeger MP remarked

> When I was in my constituency last weekend I asked my constituents ... which they would rather I did - endeavour to catch Mr. Speaker's eye in the grand foreign affairs debate tomorrow or raise the question of their bus shelter, which is only a local problem. They told me any fool can speak on foreign affairs and no doubt several would, but that if I did not speak about their local bus shelter, then nobody else would (Wallace 1975 p 95).

In addition, therefore to the formal limitations on the Parliament, there are the political imperatives of responding to constituent concerns, and the general question of the interest of voters and MSPs in international issues. It is interesting to note that the 'international' issues that have been raised in questions or touched on in debates in the Scottish Parliament, have overwhelmingly tended to have a constituent economics or job dimension. Thus questions have been raised about:

• the Seattle World Trade Organisation negotiations and their impact on the Scottish economy and local businesses; for example, a series of questions as whether the Minister had consulted Scottish Natural Heritage, the Scottish Landowners Federation, Convention of Scottish Local Authorities, Scottish CBI, the Federation of Small Businesses and representatives of Scottish registered companies etc. The answer from

Henry McLeish (Minister for Enterprise and Lifelong Learning) was that:

> The negotiation of International Trade Agreements, including World Trade Organisation talks is a reserved matter and consultations with interest groups, including industry, is a matter for the UK Parliament.
> The Scottish Executive is in regular contact with DTI on Trade issues and the WTO talks and how these may impact on the Executive's responsibilities (Scottish Parliament Official Report 13 December 1999).

While the debacle in Seattle has in the short-term rendered some of MSPs concerns moot, the impact of international disputes on trade and international dispute procedures was dramatically brought home by the protectionist conflicts between the EU and US, especially by the conflict over the EU post-colonial preference to imports of Bananas from the Caribbean. This disadvantages banana growers in Latin America, where production is in the hands of US food giants. They egged on the US government into a complaint to the World Trade Organisation, which repeatedly sustained the complaint. Early in 1999 the US produced a 'hit-list' of $320m. worth of imports from the EU which would be subject to 100% increases in duties. Particularly badly hit initially were cashmere sweater manufacturers in the Borders of Scotland, it being calculated that up to 30 manufacturers - most of them in Hawick and Innerleithen - could be forced out of business. While this was legally a problem for Her Majesty's Government in London, the EU and WTO, clearly elected local MSPs were required to become involved in lobbying and seeking solutions amenable to local concerns - if the local MSPs did not do so, then questions about their role and purpose will grow.

- The issue of the implications of the multi-lateral agreement on investment (MAI) for Scotland's trading position, to which the answer from Henry McLeish was basically the same as that above on the WTO negotiations, although he did acknowledge that anything that adversely affected Scotland's world trade position would be a matter of concern.
- Take-overs of Scottish companies, some by other British concerns, but some by overseas corporations.
- The retention of RAF Buchan.

Other questions have focused on the local and Scottish consequences of Britain's international policy, for example

- The attitude of the Scottish Executive to the presence of the Trident installations in Scotland, it being alleged that their presence was contrary to the wishes of 85% of the Scottish people, the STUC and the Scottish parties. This set of questions were made all the more poignant given that Sheriff Gimblett of Greenock Sheriff Court had acquitted three anti-nuclear protesters who allegedly caused £80 000 damage as Faslane Naval Base, which houses the British nuclear deterrent, on the grounds that she accepted the defence argument that the International Court of Justice had declared nuclear weapons were illegal. This ruling allowed Alex Salmond and others to raise the nuclear question in the Parliament.

- the issue of the impact of accommodation costs for Kosovo refugees on local councils and whether they would receive additional support to meet the additional costs, as well as associated questions about the implications for Scotland of asylum seekers.

There has also been a general awareness that Scotland's voice needed to be heard on issues such as climate change, especially following a report in December 1999 that suggested

> Rising sea levels, a 20% increase in rainfall, worsening public health and the loss of many wild species (Climate Change: Scottish Implications Scoping Study 1999).

and raised issues that clearly demonstrated the need for the co-operation between London and Edinburgh to meet international obligations and formulate positions.

The Concordats or 'Memorandum of Understanding and Supplementary Agreements'

On several occasions when some of the above issues were raised in the ScottishParliament reference was made to the Concordats, especially on European Union Policy Issues and International Relations, which were officially designed to ensure that the Scottish Executive was fully involved on all European Union and international issues which touch on devolved matters. These documents were finally unveiled in the autumn of 1999, two years after the publication of the White Paper *Scotland's Parliament* and nearly a year after the Scotland Act was passed.

Such agreements were necessary because as has been clear from the above Scotland has a clear interest in many matters that were 'reserved' and on many issues there is some blurring of responsibility between the two governments. Ambiguities result also from the fact that while the 1997 White Paper suggested how Her Majesty' Government envisaged how the new system might operate, the Scotland Act did not legislate on many of these matters and how the relationship was to actually work and evolve. In fact, the handling of day to day matters is a matter of politics and practice, precedent and convention, although within a certain framework, thus the need for concordats. The concordats attempted to set out the working practices that will apply in the relations between London and Edinburgh (with others referring to Wales), including the procedures to ensure that Scottish interests and views are taken fully into account in formulating the overall UK approach. The Memorandum of Understanding recognises that the Scottish Executive has a responsibility for ensuring that the interests of Scotland:

> In non-devolved matters are properly represented and considered

but it went on to make clear, and this has been repeated by Scottish ministers in the ScottishParliament, that the:

> Memorandum is a statement of political intent, and should not be interpreted as a binding agreement. It does not create legal obligations ... It is intended to be binding in honour only ... Concordats are not intended to be legally binding, but to serve as working documents (Scottish Executive Memorandum of Understanding and supplementary agreements 1999).

Officially central to the arrangements between the different administrations are :

- Good communication;
- The attempt not to constrain the discretion of the devolved institutions; but to seek to ensure that they have time to make representations;
- The need to respect the mutual confidentiality of discussions;
- That there must be adherence to the resultant UK line;
- That the devolved administrations are responsible for implementing international, ECHR and EU obligations;
- That most of the contact is on a bilateral or multi-lateral basis, directly between departments with most business being conducted through normal administrative channels but that;
- Some central co-ordination system was necessary. A Joint Ministerial Committee consisting of UK, Scottish and Northern Ireland ministers and

members of the Welsh cabinet has thus been created. The JMC can meet at Prime Minister/First Minister level, but also in what are termed 'functional' formats, i.e. agricultural or environmental ministers. The major meetings of the JMC are to be chaired by the Prime Minister or his deputy. It is presumed that issues will only go to the JMC when there has been a lack of success in bilateral exchanges at ministerial level.

Specifically in the 'Concordat on International Relations - Scotland' and the 'Common Annex' (shared with Wales), having reiterated the formal position making clear HMG's responsibility, there is recognition that:

> The conduct of international relations is likely to have implications for the devolved responsibilities of Scottish Ministers and that the exercise of those responsibilities is likely to have implications for international relations. This Concordat therefore reflects a mutual determination to ensure that there is close co-operation in these areas...

The Concordat thus covers:

- Exchange of information.
- Formulation of UK policy and conduct of international negotiations.
- Implementation of international obligations.
- Co-operation over legal proceedings.
- Representation overseas.
- Secondments and training co-operation.
- Visits.
- Public diplomacy, the British Council and BBC World Service.
- Trade and investment promotion.
- Diplomatic and consular relations.

In addition to these general issues, it was made clear that only the UK government could enter into legally binding treaties or other international agreements, although the devolved administrations would be allowed, in co-operation with the FCO to:

> Make arrangements or agreements with foreign national or sub-national governments or appropriate counterparts in international organisations, to facilitate co-operation between them on devolved matters

provided these did not seek to bind the UK, prejudice its interests or affect its international relations. It was also agreed that it 'may be appropriate' for Scottish ministers or officials to be part of a UK delegation, but that they would have to support the single UK line. It is the UK lead minister who retains:

> Responsibility for the negotiations and ... determine(s) how each member of the team can best contribute

and indeed determines whether a Scottish minister is to be part of the team, a matter which has already caused some difficulty in the environmental area. The devolved administrations can establish offices overseas dealing with devolved matters.

In many ways the most controversial issue proved to be the arrangements for 'Trade and inward investment', since the Scottish and UK governments have concurrent powers to promote exports and inward investment. With regard to inward investment the concern was to stop different parts of the United Kingdom engaging in competitive bidding against each other, as had happened in 1997 when there was a bidding war between the Welsh Development Agency and its north-east of England rival. Under the Concordats, those different parts were to conform to guidelines agreed by the Committee of Overseas Promotion, on which they would be represented. The key issue was who or what would adjudicate when Scotland was fighting it out with areas south of the border for grant aid on the siting of new industries, and the development or attraction of new industries. Initially it had been feared that the British Department of Trade and Industry would be the final arbiter, but in the end result this role was accorded to the Cabinet Office which was seen as more neutral. There are also to be agreed limits covering major schemes of financial assistance to companies. Both the SNP and the Conservatives attacked these arrangements, arguing in particular that Locate in Scotland would now face more restrictions on its operations than before devolution, and would waste time in fighting its corner in the COP system.

Conclusion

Scotland does have other models to follow of course, although the particular role and powers of other SNAs or NGCs varies with their own political and constitutional position. One of the most powerful models is that of the Länder in the Germany, but there is the crucial difference that the Länder operate in a federal system with a written constitution and entrenched rights. In the case of Spain the 1978 constitution assigned exclusive jurisdiction over international relations to Madrid. Hocking (1998) argues that in the period up to 1989 the central Spanish

Constitutional Court took a broad view of international relations, so that it constrained the Autonomous Communities' international ambitions, but that since 1989, it has defined international relations more narrowly, such that it is now seen as focused on war powers, conclusion of treaties, recognition of foreign governments and overseas representation. Even here, as seen above, there can be problems of definition.

The experience of other systems raises the question of whether the consultation is genuine and how much the views of SNA's and NGC's are taken into account. While lawyers may pay attention in formal models of foreign policy, political analysis requires the examination of practical implications and other processes, often informal. In relations between the London and Edinburgh administrations they are still finding their way. The real test will come when the political complexion of the regimes in Edinburgh and London are significantly different.

The new Scottish Parliament and Executive will not only be dealing with HMG, the BIC and EU and other macro-international institutions. In order to maximise Scotland's influence, they will wish to establish links other regions and even states. This will raise a number of issues about the strategic basis of such links and the role of each body in pursuing and maintaining them. An example of the sort of contact and issues involved can already be seen since in recent years as Scotland has already forged links with a number of EU (Sweden and Finland) and non-EU states (Norway and Iceland). These links, which in Scotland's case have focused largely upon the Highlands and Islands, are based upon common features (e.g. geographical peripherality) and socio-economic interests (e.g. regional development). Four areas of co-operation have been pursued (IT, SMEs, forestry and further education) with informal links having been developed in the environment area as well. Further co-operation on a range of other issues will surely follow (e.g. rural development). Many devolved matters thus now have not only an EU but also broader European and international dimension.

The foregoing has demonstrated the ubiquity of international relations, and how wide-ranging their impact can be on Scotland. For the Parliament, is also means that committees like 'Enterprise and Lifelong Learning', 'Rural Affairs' and 'Transport and Environment' will have a strong international dimension.

In an era of globalisation, the local is increasingly international and vice versa. That is, the boundaries delineating territorially defined political arenas are becoming increasingly fluid (Rosenau 1997). As Hocking has observed the notion that there is an accepted and uncontested hierarchy of interface with the international arena is challenged as SNAs or NGCs of various kinds respond to forces generated by domestic and international change (Hocking 1993). The attempts by central governments to assert control over a fragmenting international political environment whilst at the same time recognising that new policy agenda create mutual dependencies between actors, both public and private which demand constructive management strategies, which have clearly not been fully confronted in the UK/Scottish case. The longer-term answer will depend up how the practice of the Concordats evolves but it will also be crucially dependent on the aspirations of the Scottish people and their representatives.

References

Aguirre, A. (1999), 'Making Sense of Paradiplomacy? An Intertextual Inquiry about a Concept in Search of a Definition', in Francisco Aldecoa and Michael Keating (1999) (eds), *Paradiplomacy in Action:The Foreign Relations of Subnational Governments*, Cass, London.

Climate Change: Scottish Implications Scoping Study (1999), quoted by *The Scotsman* 3 December 1999.

HMSO, *Scotland's Parliament* (Cm. 3658, July 1997).

HMSO, The Scotland Act.

HMSO, Scottish Parliament-Official Report.

HMSO, House of Common Session 1997-98 Scottish Affairs Committee.
The Operation of Multi-Layer Democracy Minutes of Evidence 25 February 1998.

HMSO, The Agreement.

Hocking, B. (1993), *Localizing Foreign Policy : Non-Central Governments and Multi-layered Diplomacy*, Macmillan, London.

Hocking, B. (1998), *Foreign Policy and Devolution, Conference Paper, British International Studies Association*, 24th Annual Conference December 1999.

Jones, R. (1970), *Analysing Foreign Policy: an introduction to some conceptual problems*, Routledge and Kegan Paul, London.

Keating, M. (1992), The Rise of the Continental Meso : Regions in the European Community in Sharp, L.J. (ed), *The Rise of Meso Government in Europe*, Sage, London.

Keating, M. (1999), Francisco Aldecoa and Michael Keating (eds), *Paradiplomacy in Action: The Foreign Relations of Subnational Governments*, Cass, London.

Kissinger, H. (1969), 'Domestic Structure and Foreign Policy', in Rosenau J. ed *International Politics and Foreign Policy*, Free Press, New York.

Marks, G. (1992), 'Structural policy in the European Community', in Sbragia, A (ed). *Euro-Politics: Institutions and Policy-making in the new European Community*, Brookings Institution, Washington DC.

Marks, G. (1993), 'Structural policy and multi-level governance', in Calfuny, A. and Rosenthal,G. eds *The State of the European Community Vol. 2: The Maastricht Debates and Beyond*, Longmans, London.

Mosely, Philip E. (1961), *Research on Foreign Policy in Brookings Dedication Lectures*, Research for Public Policy, Brookings Institution, Washington DC.

Rosenau, J. (1980), *The Scientific Study of Foreign Policy*, Pinter, London.

Rosenau, J. (1997), *Along the Domestic-Foreign Frontier: Exploring Governance in a Turbulent World*, Cambridge University Press, Cambridge.

The Scotsman
The Sunday Herald
Wallace, W. (1971), *Foreign Policy and the Political Process*, Macmillan, London.
Wallace, W. (1975), *The Foreign Policy Process in Britain*, Royal Institute of
       International Affairs, London.

PART V
PRESSURE GROUPS
AND
CIVIC SOCIETY

12 Redemocratizing Scotland. Towards the Politics of Disappointment?

GRANT JORDAN AND LINDA STEVENSON

This chapter looks at one aspect of the delivery of Devolution: the place of interest group representation in what has been heralded as a New Era in Scottish politics. If we are in such an epoch, do interest groups add to, or detract from, the quality of democracy? Some commentators believe that in Scotland's 'New Democracy' there will be no need for pressure/interest groups, as *civic society* will play a greater role in policy making; although others consider that such groups undermine party based representation.

Devolution has possibly produced two main areas of change. For many observers the call for change was simply a rejection of the link to the Westminster political system and what was considered to be British domination over Scottish political preferences. But the reformers were also rejecting Westminster type politics, as well as Westminster itself. In this light the Scottish Parliament was intended to be more than just the old regime writ small: it represented a chance to avoid the flaws in the British system, and to move away from the confrontational style of politics, to embrace a new, and arguably more democratic, way. Broader participation and consensus was to be the remedy. The influential document from the Scottish Constitutional Convention (SCC) (*Scotland's Parliament, Scotland's Right*) declared that its own policy making process (involving potentially conflicting participants) had proved that, '... constructive consensus is achievable, even among those steeped in the ritual confrontations of British politics. That lesson is immensely encouraging [they claimed], not just for the project of designing a Scottish Parliament, but for the much more important question of how the Parliament will work once it is in place ...' They went on to say that they saw that '*the consensus that this report represents [is] a beacon of hope* for a new and better politics in a Scotland running its own affairs' (italics added). The document claimed that the protracted discussion of the Constitutional Convention, with no majority decision making on major decisions, was, 'A vision of a new kind of politics, one in which we have come to see great value.' The report argued for a Scottish Parliament 'to usher in a way of politics that is radically *different* from the rituals of Westminster: *more participative, more creative, less needlessly confrontational* (italics added).'[1]

By October 1999 significant developments had taken place in connection with the implementation of some kind of mechanism to give effect to these ambitions for an integration of 'civic groups'. This chapter seeks both to describe these developments, to note some contradictions within the largely uncritical support for them, and to query the thrust of the reform agenda.

Scottish Civic Forum

Prior to the 1999 Scottish elections, interest groups within Scotland were mobilising their resources in order to focus on the new Parliament. UK-wide groups also further developed distinct Scottish offices, one example being the CBI Scotland, which has more autonomy as a result of devolution. Situations vacant advertisements in the press contained numerous adverts for parliamentary liaison officers for public, private and voluntary sector organisations desperate to ensure that they would not be left behind in the race for inclusion. Informal networks of organisations discussed how best they could work together to ensure that the apparent consensus in favour of group participation in the pre-legislative process would not fade once politicians, and partisan interests, became involved.

But the major development in the move towards wider participation is the emergence of the Scottish Civic Forum (replacing the Scottish Civic Assembly). This seeks membership from religious and business organisations, trade unions, voluntary organisations and professional associations[2]. The aims state that, 'Our vision is that the Scottish Civic Forum will break the mould of old fashioned politics. It will increase participation, find new ways to open up dialogue, raise awareness and stimulate debate on the many challenges facing Scotland.' A major principle of the Forum is that, 'The Scottish Parliament should embody and reflect the sharing of power between the people of Scotland, the legislators and the Scottish Executive.' The aims asserted that it will harness 'the experience, practical expertise and capacity of Scottish *civic life* … It will be an avenue for finding new ways to increase dialogue and understanding between the people of Scotland and their Parliament.'[3] The aims also set out that the Forum will be 'an additional way of getting the people of Scotland involved in the work of the Parliament - it will not replace existing bodies or relationships, nor will it stand in the way of direct representation … It will adopt policies of active participation from otherwise excluded groups and unheard voices.   (It) will explore the prospects for consensus across civic life but will also reflect and record disagreement and diversity where they exist. It will not contrive consensus where none can be agreed.' The recommendations therefore laid stress on consultation, suggesting that, 'Draft Bills on Introduction should be accompanied by a memorandum explaining the consultative process undertaken and the degree of consensus reached, as well as …the consultative process undertaken.'[4]

The Report of the Consultative Steering Group was debated by the Parliament on 9 June and, there was much (uncritical and conceivably superficial) support for the Forum. Just as there had been prior to the election, there was wide cross-party support for its establishment, with calls for financial assistance from within the Parliament to encourage broad participation. Hopes were expressed, by George Reid for example, that a Civic Forum would promote participation, facilitate debate, and ensure social partnership, and that it would be a 'gateway to the Parliament and its Executive, not a gate-keeper.' This was supported by Karen Gillon who saw the Civic Forum as a way of ensuring a Parliament for all Scotland. According to Cathy Peattie, the 600 groups that had registered an interest in joining the Civic Forum would permit dialogue to be conducted across the civic sector, and she suggested that its establishment should not be seen as a threat to Parliament, but as being complementary to it. Since then more than £300,000 has been allocated by the Scottish Executive towards the running of the Forum, over a period of three years, and promises have been made for further assistance in kind.

Rejecting Parties: Reforming Democracy?

Developments concerned with the transformation of democratic practices have emerged because many advocates of Devolution were at least subconsciously dissatisfied with representative, party - based politics[5]. 'Confrontational' was the preferred (damning) label for rejected practices. The image involved was of adversarial extremism in which Governments (often polling well under half the votes) *imposed* policies on Scottish society. This sort of complaint found voice in Scotland due to the prolonged phenomenon of a Westminster Conservative Government controlling a Scottish Administration. The critique conflated the ideas that elections were trivialised - ignoring big issues - and that elections produced ideologically extreme swings.[6] Associated with this rejection of the representative/party model is the apparent decline in mass partisanship and party mobilisation implying that the electorate has a cynical view of politics and that there is a need to find a way to reduce voter apathy: to engage the wider public in politics.[7]

This discussion of the reform of Scottish democracy connects to contemporary discussions within political philosophy about democratic reform of decision making. Dahl (1989, p106) points out that the literal meaning of democracy is 'rule by the people', but in order to rule there must means to permit this. His (1956, p3) general, and minimal, definition of democracy is *'processes by which ordinary citizens exert a relatively high degree of control over leaders'*. While his basic notion of a democracy is, 'the continuing responsiveness of the government to the preferences of its citizens, considered as political equals' (1971, p2), *how* that can be achieved is not self-evident. Different forms of democracy exist.

Periodically views have changed over whether democracy is/should be *participatory* and concerned with self-government and self-regulation - or whether it is *representative* with participation restricted to the choice of leaders who have the authority to make decisions. Conventionally there are two main categories for discussion,

> direct participatory democracy,
> and representative or liberal democracy.

Democracy (i) The Direct Participatory Model

For much of recorded history, democracy has meant *participation in decision making* rather than (merely) the *selection of decision-makers*. The direct participatory approach has been revisited in recent decades in different terms for example: deliberative democracy, and developmental democracy. It is such devices that have regained popularity. So the reformed Scottish system, advocates say, should not just be Scottish, but participatory.[8]

Dahl (1989, p13), termed the emergence of citizen involvement in decisions as the First Democratic Transformation. He sets out that the associated implied pre conditions of this *ideal* (practice was probably far less satisfactory) were at least two-fold. Firstly, the society was not expected to contain unbridgeable differences in wealth or attitude, or lines of conflict so severe that no common policy was feasible. Secondly, the political unit had to be of modest size. Hudson (1995, p5) set out that, 'Only in a small state, where people could meet together in the relative intimacy of a single assembly and where a similarity of culture and interests united them, could individuals discuss and find the public good.' The smaller scale of a Scottish democracy *might* be seen as making participatory democracy more practical.[9]

In his account of the participatory ideals dominant in classical Athens and Rome and in the city states of Italy, Held (1996, p89) made the main point that democracy was thought to rest on *active citizens*. The nature of democracy in classical times is relevant not simply because of its historic importance in shaping the debate about democracy, but because it is this core idea of the active individual that is at the heart of contemporary accounts of democracy. Held says that in Renaissance republicanism, as well as Greek democratic thought, a citizen was someone who (to quote Aristotle) participated in, 'giving judgement and holding office'. Held maintains that the limited scope in contemporary politics for the active involvement of citizens would, classically, have been regarded as undemocratic. He points out from that perspective there is great difficulty in locating citizens in modern democracies, except as elected representatives or office holders.

In the decision-making meetings in the Greek polis, the aim was not to seek to pursue one's own interests, but to discover the interests of the whole: that was why face to face interaction was so important. This interpretation of participatory democracy was set out by Mabbot (1958, p30) - writing about Rousseau - who said that citizens are called together to vote *not for what each wishes* but *each for what all wish*. He emphasised that for Rousseau the 'general will' is different from the 'will of all' which is simply a mere total of selfish wills. The pursuit of individual self-interest is rejected. While many would seek to restore the importance of individual participation as a requirement of democracy, there is often a lack of clarity about the basis of that participation. Is the idea that individuals try to get their own way (as in a referendum or an anti motorway protest)? Or do they try to establish a common and collective preference, as in the sense advocated by Rousseau? This is one (of the numerous) tensions in the current agenda of those seeking the redemocratization of Scotland.

Democracy (ii) Representative Democracy.

The initial formulation of democracy was then individual and activity based. The key to the revolutionary idea of *representative democracy* was that members of the public could exert a very broad control of the key personnel and policy outputs of the system - without investing their own time in policy making activities in which they could (because of the scale of the numbers of individual potentially involved) expect to have little noticeable impact (i.e. their actions are unlikely to be pivotal). It sought to deliver to the citizen the benefits of direct participatory democracy (i.e. responsiveness of the system to popular wishes) by a very different (less demanding, but arguably more realistic) arrangement. The citizen secured satisfactory outcomes by *determining his or her representative*. The choice of representatives was an indirect means of selecting outcomes.

Thus as political units increased in scale through the centuries, the direct formulation of democracy seemed less and less practicable. As Dahl (1989, p328) noted, 'Most people took it as a matter of self-evident good sense that the idea of applying [the individual - centred version of the] democratic process to the government of the nation-state was foolish and unrealistic.' Accordingly practical reasons saw the requirement that public participation be accommodated by the device of representative government. A reformulation of the notion of democracy, which accepted the representative principle, appears to be needed if the larger scale 'nation states' (often in fact subsuming several nations) that developed could be considered democratic. The definition of democracy had to alter to fit the political societies that wanted to make a claim to that status. Dahl (1989, p213) terms this refashioning of the principle to fit the requirements of new units as the Second Democratic Transformation. He thus presents democratic theory not as a continuously evolving interpretation, but as a pattern of radically different, and indeed, antagonistic, ideas. The representative idea was a contradiction of the participatory. As in the Greek tradition, the new democracy was legitimated by public participation - but this was now a limited type of participation confined to electoral choice.

Democracy (iii) Group-based, Pluralist Democracy

However having evolved a theory of representative democracy that assumed that best policy emerged from the aggregation of votes, political science found that turnout was often low and that those active in voting were rarely well informed to satisfactorily make even basic decisions such as candidate choice. There is a third group-based interpretation of democracy in which, it is argued, individuals *qua* individuals can have a limited role, but they can act effectively through interest groups. This view about the importance of groups in policy making is well established, but groups lack a prominent place in democratic theory. Because much discussion has been informed by the basic twin - track direct /representative formulations of democracy, there is a tendency to assume that any alternative to representative democracy (such as group activity) thereby relates to direct democracy. This assumption has to be queried. Group democracy is important, but not necessarily because of the participatory opportunities it gives individuals. Indeed as we have already noted, historically the stress has been on the role of the individual: the group was seen as a threat to democracy. Modern empirical democratic theory (aka pluralism) accepts a truncated and minimal participation by individuals. In this version, as Dunleavy and O'Leary (1987, p23) observe, 'Pluralists know that citizens do not *and cannot* directly control policy making in polyarchies.' They quote Dahl as remarking that politics for most citizens is a 'remote, alien and unrewarding activity.' In similar vein, Polsby (1980, p117) commented, 'If a man's major life work is banking, the pluralist presumes that he will spend his time at the bank, and not in manipulating community decisions.' This idea of the political sphere as simply an option in life is extended to what others might see as political 'apathy.' The pluralist account accepts distinctively limited political appetites among the public and makes few demands on them. Democratic control comes about through a combination of competitive elections and interest group representation on behalf of sectional interests.

The major inconsistency in the reform agenda is the welcome given to pluralist-like group participation *and* the stress in participatory democracy to individuals with activity beyond simple voting. Arguments about involving *major* groups in some kind of corporatism (in a very broad, non - technical sense) in which they share a decision making role with elected Government, is very different from the Civic Reform agenda model that sees small groups and individuals influencing Parliament. (The tension between the place of groups and individuals is not explicit in the reform package. A system that enhances civic group access may diminish the influence of individuals.) In the past few decades an active citizenry has re-emerged as an ideal: accordingly representative democracy has for some time been seen as an inadequate version of democracy. This stress on personal activity means that current thinking differs from pluralist group democracy. Carole Pateman's *Participation and Democratic Theory* (1970), and Benjamin Barber's *Strong Democracy* (1984), reject the sort of minimal democracy tolerated by pluralists: the call was for more '*authentic democratic politics*' - not only in the political realm but in the workplace and other institutions (Hudson, 1995, p20).

Civil Society and Deliberative Democracy: 'A Better Way'

The term Civic Forum invokes the notion of civil society. As Walzer describes (Mouffe, 1992, p89) this has (re) emerged to describe moves in post Communist societies to revive non-state institutions. It involves rebuilding 'networks: unions, churches, political parties movements, co-operatives, neighbourhoods, schools of thought, societies for promoting and preventing this and that.' But the term Civil Society is positively 'loaded'. And hence borrowing it smuggles in connotations suggesting that the call to give certain social groups more weight has the same kind of virtue as dismantling authoritarian regimes[10].

The concept of civil society is central to the theory of deliberative politics. Civil society, an idea that can be traced back to Fergusson, is also a central term in the reform agenda in Scotland, but it is far from clear if the term (or the 'civic' variant) is being used in the specialist senses of political philosophy. [11] If civil society is seen to be the institutional framework of a modern society then, according to Cohen and Arato (in Axtmann, 1996, p72),

> It is constituted by three complexes of rights: those securing the socialization of the individual (protection of privacy, intimacy and the inviolability of the person); those concerning cultural reproduction (freedoms of thought, press, speech, and communication); and those ensuring social integration (freedom of association and assembly).

The actions of individuals in such a society are bound up in an awareness of what is expected of them in virtue of the social bonds that tie them each to the other. In deliberative democracies collective action is taken to influence the political and economic subsystems. There is a willingness to embark on an open-ended process of discursive will formation that involves a willingness to reverse perspectives and to reason from the other's (others') point of view (in Axtmann, 1996, p74).

They go on to say that political society in the form of representative democracy and civil society together provide the necessary elements that permit deliberative politics to be pursued. They highlight the complex interdependencies that exist between political society, civil society and the private lifeworld. However, as Axtmann (1996, p76) points out, Habermas argues that the public opinion that is worked up via democratic procedures into communicative power cannot 'rule' itself, but can only point the use of administrative power in specific directions. Actors in civil society have no power to make decisions, only to influence the decision-makers.

While there is an extensive academic discussion of civil society, the use of the term in Scotland may be no more than an assertion of the merits of more consensual decision-making[12]. One instrument towards that end was an electoral system that would produce greater proportionality among the elected. This is presented as in itself a 'good thing', but the implication was that this proportionality would frustrate 'elective dictatorship' by majoritarian governments resting on a minority of electoral votes.' The current stress on civic participation can be seen as a remedy to the pre Devolution situation of political rule that was often out of step with Scottish electoral opinion. Consistent with that vein of thinking would have been the emergence of a minority Administration negotiating ad hoc majorities on different policies. In the event, the prospect of a minority administration was seen as some kind of failure by the media, and a coalition between Labour and the Liberal Democrats was treated as a 'success'. However, this 'success' may frustrate the broad consensus-building aspect of the reform. Once it was decided to go for the convenience of a coalition rather than a minority administration, then the likelihood of much significant negotiation within the Parliament, and between forces in Parliament and the wider society, was curtailed. To be blunt, once politics became a matter of for and against the secure Executive, then the focus has to be on the civil servants and politicians in the Executive, rather than on civil society and the Parliament.

The Final Scottish Constitutional Convention Report declared that the new Scottish Parliament would 'differ from Westminster in a less procedural, and more radical, sense.' The Scottish model, as presented by the SCC, would ensure it was responsive to the wishes and values of the Scottish people by taking the 'views and advice of many specialist organisations and individuals.' The proposed Standing Orders provided for electors to 'directly petition the parliament; required the legislators to consult widely both before and during the legislative process; would provide facilities for the public and the media to meet MSPs easily; and aimed to encourage and promote constructive rather than confrontational debate and discussion. The SCC Report ends with ambitious rhetoric,

> We offer not just a new parliament, but the possibility of a renewed nation, a cathedral of hope and promise for all the people of Scotland, and indeed the whole United Kingdom ...We are citizens not subjects. We are partners not customers ... We deserve something better than the secretive, centralised, self-serving state that the UK has become.

In the Second Reading of the Scotland Bill Donald Dewar endorsed the idea that the reformed process would deliver a superior sort of democracy, 'The Bill will be welcomed by democrats everywhere - it is not simply about Scotland, nor is it in any sense routine reform tinkering with the detail of our political system. It goes to the heart of our democracy and offers hope for the democratic process itself.' He said that the change implied 'a new flexibility, new thinking, and perhaps new forms of cooperation'. He went on to say that a new balanced partnership between the executive and legislative branches of the state in Scotland was being sought that would promote consensus and good government. There would be far more give and take; more listening; and more cooperation.

The Consultative Steering Group on the Scottish Parliament set up as an advisory body elaborated further on the particulars of policy development. This (along with other initiatives such as petitions and citizens juries) laid stress on the work of Parliamentary committees as a conduit for extra parliamentary opinion; they would have the power to take oral and written evidence. It argued that prelegislative involvement by Committees - 'a capacity to conduct inquiries which would include the taking of written and oral evidence' - would 'allow individuals and groups to influence the policy- making process ... By making the system more participative, it is intended that better legislation should result.' It was argued that seeking comments on specific legislative proposals did not meet the CSG's aspirations for a participative policy development process - that currently it was too difficult to affect detailed legislative proposals, 'What is desired is an earlier involvement of relevant bodies from the outset - identifying issues which need to be addressed, contributing to the policy- making process and the preparation of legislation.' As will be argued below, this kind of discussion underestimates the volume of consultation that has been typically conducted by the Scottish Office and Westminster Departments. The report also claimed that it was difficult for interested groups and individuals to keep track of secondary legislation, and that consultation on secondary legislation was limited. This was an assertion with no supporting evidence.

The CSG Consultation Paper called for a sharing of power between the people of Scotland, the legislators and the Scottish Executive,' and it asked for comment about the arrangements for involving civic society, including women's groups, people from ethnic minority communities, people with disabilities, business and the general public. The Analysis of the Responses to the [CSG] Consultation Document on the Operation of the Scottish Parliament claimed that there was widespread support for any mechanisms which could be found to involve the public, civic society and representative groups in the work of the Parliament. The Report stated that,

> In particular, voluntary organisations and interest groups were particularly keen to see that formal systems were put in place to allow them to speak as representatives of their constituent groups. Many voluntary organisations in particular (for example, Fair Play) suggested that the Parliament should set up a database of registered consultees who would automatically be approached on any relevant issues. Voluntary organisations were very keen to see that a distinctive role was given to them acting as a conduit between government and civic society. For example, the Royal Society for the Protection of Birds stressed that non-Governmental   Organisations already involved a wide variety of the population who felt that such interest groups more accurately reflected their concerns than the traditional political parties. It would therefore be sensible to use such NGOs and their existing networks (and the basic

consultation mechanisms already in place) to encourage greater consultation with, and participation by, the public in the political process.[13]

Several comments can be made at this point. Firstly, there is an element of the Mandy Rice Davies, 'He would, wouldn't he' here. If organisations are asked if they would like more influence, the answer is predictable. Secondly, many comments underestimated the ease of access to 'old style' consultation procedures. Thirdly, it is true that organisations such as the RSPB do have large memberships but it is far less clear that those joining do so for political reasons. It is most unlikely that the membership expects to see the RSPB commenting on topics such as health and education. And fourthly, while it is one thing to argue for a prominent participatory place for the 'big number' groups such as the RSPB, or trades unions, or whatever, the recent fashion for inclusion seems to want to privilege very small groups. If the small groups are relabelled as 'not popular' the arguments sound less compelling.

The basis on which some might want to involve large groups has to be different from the argument about small groups. Why should groups with very few members be accorded much of a stake? Or why should individuals with a narrow expertise be listened to outside that narrow terrain? This is not to say that there is anything wrong in allowing small groups access, but it is saying that expectations should be modest. There must be huge weaknesses in our system of policy development if there are important changes to be made to any policy on the basis of their comments. It is the responsibility of civil servants, perhaps the major national organisations, parties and the full time politicians to make policy. This suggests some restraint on expectations about the impact of the micro organisations; citizens who are not active should not have less control over public policy processes than the tiny fragment who are mobilised in small groups. There is a potential here for a new bias.[14]

The CSG Consultation Report implied, but very negatively, that there was already consultation. This was disparagingly referred to as consultation only with the '*usual suspects*'. The implication is that some interest groups are overpowerful but there are *socially* excluded groups who are also *politically* excluded. The prescription seems to be that these groups can be empowered by new procedures. RNIB suggested Braille consultation materials: Grampian Racial Equality Council suggested interpreters at public consultation meetings; WEA suggested a programme of education for all citizens. In short there were calls to encourage and facilitate participation by minorities. This is not to say that in a democracy the views of minorities should be discounted but what might be unwise is the assumption that minorities are the repository of worthwhile views simply because they are small scale. Nor should we assume that the views of the few (often very conservative in their outlook) who turn out at public meetings should receive more weight than those who are prepared to participate through the ballot box. To object to reforms to enhance the input of individuals and small groups smacks of an attitude matching Scrooge's distaste for Christmas, but it could well be that the active minority will be advantaged over those doing equally useful things in society such as running homes, or participating on other apolitical organisations. The new 'inclusiveness' runs the danger of being a cover for increasing the influence of self selected activists who do not require to test their views against the electorate.

The Diagnosis: Second Opinion Needed?

The new agenda is about pre-legislative discussion, public petitions, citizen juries, wider access. There is a fairly clear picture of a widespread acceptance of a deficiency (lack of consultation) in Scottish political life - a complaint for which the Scottish Civic Forum appears to be the remedy. But both the diagnosis and the prescription deserve some critical attention. The readiness with which even legislators backed the Civic Forum indicates something of a lack of confidence in the principles of representative democracy.

Essentially there is a need to question the assumption that the new ingredient required for British and Scottish policymaking is *more* consultation. In fact pre-devolution policies within Britain characteristically arrived at by consultation and negotiations between Whitehall and relevant interests. Getting involved in the consultation process is not as difficult as critics claim. Consultation has been the departmental routine throughout the 1980s and 1990s at least - even where there was no legislative compulsion to do so. As one DOE civil servant commented, 'consultation is now part of the scene and is unlikely to go away'. A Home Office civil servant observed, 'consultation occurs because the policy outputs need to be implemented, and because the 'right' outputs need to be achieved.' Extensive consultation is an integral part of the 'management of pressure' (see Jordan and Richardson, 1987).

In July 1998 at a time when agriculture's relations with Government were strained by BSE, strength of pound and other factors the National Farmers Union of Scotland was still reporting to members (via the *Scottish Farming Leader*) that over recent months they had been involved in consultation on topics such as:

Access Policy, Animal Medicines, Cattle Database Regulations, Farm Business Incomes, Groundwater, Land Reform, Milk Pricing, Rural Development Strategy, Sheep Carcass Exports. Working Time Directive, Agenda 2000, Calf Processing Aid Scheme, Data Export Scheme, Food Standards Agency, Industry Restructuring (Retirement), Merger of Agri - Environment Schemes, Pig Identification, Seed Potato Classification, Sheep Welfare Code.

The use of consultation is generally defended as a means to improve policy or even to establish policy. The policy relevant group or even firm can simplify the policy making task for the civil servant. They may be actively pursued because they are the possessors of 'indispensable information' which decision-makers lack. They see consultation as a necessary part of their existence. The right to consultation, even when not formally conceded by the government, is such an element in practice that any failure in the normal procedure is treated as an injustice by the group concerned.

The reformer's case is crucially fudged over who the 'participators' are to be. Within the same 'bed' are very different attitudes to the power of elected politicians, individuals, major interest groups and less influential groups. It may turn out, as so often is the case, that the banner of 'reform' and 'improvement' conceals mutually antagonistic agendas. Within democratic debate there is a major divide between 'realists' who settle for group based participation and those who are more ambitious in their expectations and seek a deliberative democracy based on individuals. Currently it seems to be assumed that anything is better than the politics of the past, but it is not against the spirit of improving democracy to be sceptical of practices that will simply undermine respect for other forms of democracy - while creating undeliverable expectations that everyone can win prizes. Democracy can be seen as a process that tends to satisfy majorities while respecting the intensely held preferences of minorities, but there is nothing 'undemocratic' about special interests being disappointed. Where reformers believe that politicians should be making final decisions *and* that 'usual suspect' groups are too powerful, *and* that participation can be created that is substantially different from 'one - side consultation', then there is much room for confusion and disappointment.

The Scottish Civic Forum website points out that,

> There are many thousands of civic organisations in Scotland operating in every area of public life. It follows that any management and governance arrangements for the Forum must reflect that diversity. All of civic society, no matter how small or remote, must be able to have their say.

But these are often groups that cannot have much claim of expertise to affect national policy. Groups that have a valuable role are likely to be well established in their policy niche. The form of words - 'have their say' raises expectations that it is fair and possible that policy can be made by small organisations.

Some believe that Civil Society is no more than an intellectual fad. Eberly comments that, 'In contemporary American society enthusiasm for paradigmatic shifts are brief. An instant sense of expectation was raised previously with such terms as 'empowerment,' 'reinventing government,' 'public-private partnerships,' and, more recently, 'the new citizenship' (Eberley, 1998, p66). Civil Society has been attacked from across the political spectrum, seen by some as 'an attempt to restrain discourse and stifle discontent ... a last ditch attempt by the old guard establishment to retain control of the social debate' (Paglia in Eberly, 1998, p67). This view is endorsed by Cohen who asserts that 'this conceptualisation of civil society ... when combined with the discourse of civil and moral decline, undermines democracy instead of making it work, threatens personal liberty instead of enhancing it, and blocks social justice and social solidarity instead of furthering them' (Cohen in Eberly, 1998, pp67-68). In this cynical light the Civic Forum and over optimistic expectations of public participation can be viewed as potentially damaging to the objective of creating a better democracy in Scotland.

The discourse about consultation assumes that what has been wrong in the past is the failure to include the views of minorities. This chapter instead assumes that the major weakness was the lack of a political and parliamentary majority to reflect Scottish political views: the democratic deficit in Scotland. Post Devolution, policy making will be that much more satisfactory when decisions can be linked to public preferences through the ballot box. In that sense the need for even wider consultation may be abated because the proposals from the Executive should be consistent with public opinion as expressed through elections. While the language of People and Parliament has a decent ring to it, the necessary reality of contemporary policy making may well continue to be Groups and Government. What we need to beware of is the politics of disappointment that may result from raising expectations beyond a realistically attainable level. The New Democracy already instituted should not be judged as failing simply because it does not meet some expectations that were, perhaps, unrealistic.

Notes

1   However the groups and individuals represented in the Constitutional Convention were united by a common goal: Devolution: and against a common 'enemy' - the UK Conservative government. The opponents of devolution, i.e. the Nationalists and the Unionists, did not participate in the debate so reducing the likelihood for major conflict, making consensus easier to achieve. This was a less than compelling example of consensus building. Quote taken from the SCC report on the Campaign for a Scottish Parliament website at http://www.cybersurf.co.uk/csoparl/briefing/scc_prop.html

2   Excluded from membership of the Civic Forum are statutory bodies, local authorities, political parties, government agencies, and for-profit organisations. As was argued, however, membership will be open to national organisations and sub-national organisations which are not local branches of national organisations (Bonney, 1999, p14).

3   Those in favour of the Civic Forum recognise that in order for such broad participation to work there must be a change in the culture of politics in Scotland. 'However, institutional arrangements alone cannot secure a fundamental and lasting change to the culture of politics. A shift in attitude and approach to the business of politics and government is required from those that work in the system, from those that report in the media and indeed from those involved in the organisations that make up wider civic society.' The working paper, *Accessing the new Scottish Parliament: changing the culture of politics*, written by an informal network of organisations that has met frequently in Edinburgh during Parliament's first year discusses ways of ensuring access is provided to groups, in keeping with the principles of the CSG report.

4   The paper referred to in note 3 makes a distinction between pressure groups and other bodies. Pressure groups, apparently, 'are more likely to use the techniques of pressure and lobbying developed at Westminster and it seems superfluous to address their needs by guaranteeing them access to the formal process of policy making ... In short, there is a need to guarantee institutional access to bodies representative of major forces in Scottish society. These would be local authorities, COSLA, the STUC, CBI, SCVO, professional organisations representative of specialisms in the public sector and the voluntary sector, ...' In fact it is hard not to think that such bodies can use existing opportunities as well as any 'pressure groups'.

5   The discussions around Devolution in the late 70's did not have this anti party vein. See, for example, McKay (1979).

6   Mae, (in Shafer, 1998), argues that the 'least partisan wins the Presidency'. This supports earlier work. In the UK it may be that Thatcherism prevailed not because of the virtue of clearly espoused policy positions, but because of fragmented opposition.

7   Martin Sime, Director of the SCVO, has been at the forefront of the campaign to introduce a Civic Forum, and has voiced support for teledemocracy, as well as for other suggested innovations that would enable the people of Scotland to participate in Scottish politics.

8   Bonney (1999, pp 5-9) points out that the new Parliament actually has the potential to enhance representative democracy in terms of the degree to which it allows scrutiny of the executive; in that it has a greater number of representatives

responsible for fewer policy areas, and a strong committee system that allows MSPs to specialise and to influence pre-legislative policy formulation. However, he does not believe that this potential is being exploited to the full.

9   But 'small states' may have been much smaller than 5m. The participatory case favoured by advocates of the Civic Forum echo the unitary (as opposed to adversary) type of politics as described by Mansbridge J (1983).

10   In fact Walzer says 'we have lived in civil society for many years without knowing it.'

11   There appears to be some confusion over the use of the term 'civic' society in Scotland. The literature of participatory or deliberative democracy refers to 'civil' society. And even in *Civic Voice*, the newsletter of the Scottish Civic Assembly, an article inviting members to join the UN Link 2000 reform programme, uses the term 'civil' society. In defining the term 'civil' society, it quotes the UN suggestion that 'non-governmental organisations; other civil society organisations and institutions such as trade unions and business and professional associations; religious, academic and indigenous entities; local administrations and the media' should all be included. The article goes on to suggest that people are members of 'civil' society 'through participation in one or more voluntary organisations or other non-state, non-governmental and non-party political entities' (The *Civic Voice*, Issue No 3, Summer 1999). However, the Scottish Civic Forum and those involved with it, refer only to 'civic' society, and it is unclear whether the term can be used interchangeably with that of 'civil' society, or whether a distinction should be drawn.

12   While the label seems borrowed from Civil Society, the sentiment may be nearer the New Left 'participatory democracy' of the 1960s (See Mansbridge, 1983, pviii).

13   Annexe D, Report of the Consultative Steering Group, p 1.

14   Also some evidence of geographical bias in mobilisation.

References

Aims of the Scottish Civic Forum, Civic Forum website
        http://www.civicforum.org.uk
Axtmann, R. (1996), *Liberal Democracy into the twenty-first century:*
        Globalization, integration and the nation-state, Manchester University Press,
        Manchester.
Bonney, N. (1999), *Scotland's New Democracy: the potential and the*
        Challenges of devolution, a paper in the Professorial Lecture Series, Napier
        University.
Civic Voice, Newsletter of the Scottish Civic Assembly, Issue No 3,
        Summer, 1999.
Dahl, R. A. (1956), *A Preface to Democratic Theory*, University of Chicago,
        University of Chicago Press, Chicago.
Dahl, R. A. (1989), *Democracy and its Critics*, Yale Press, London.
Dunleavy, P. and O'Leary, B. (1987), *Theories of the State: The Politics of*
        Liberal Democracy, Macmillan Education Ltd, Basingstoke.
Eberly, D. (1998), 'Civil Society: attacked from all comers,' in *The*

Responsive Community, vol 8, Issue 4.
Fishkin, J. S. and Luskin R C. (1999), 'The Quest for Deliberative
  Democracy,' in *The Good Society*, vol. 9, no. 1.
Held, D. (1996) (2nd edition), *Models of Democracy*, Polity Press, Oxford.
Hudson, W. E. (1995), *American Democracy in Peril: Seven Challenges to
  America's Future*, Chatham House, Chatham, New Jersey.
Jordan, A. G. and Richardson, J. J. (1987), *Government and Pressure Groups
  in Britain*, Oxford University Press, Oxford.
Mabbot, J. D. (1956), *The State and the Citizen*, Arrow Books, London.
Mackay, D. (1979), *Scotland: The Framework for Change*, Paul Harris Publishing,
  Edinburgh.
Mansbridge, J. (1983), *Beyond Adversary Democracy*, the Chicago University
  Press, Chicago.
Mouffe, C. (1992), *Dimensions of Radical democracy: Pluralism,
  Citizenship, Community*, Verso, London
Polsby, N. W. (1980), *Community Power and Political Theory*, Yale University
  Press, New Haven.
Report of the C S G (1998), http://www.scotland.gov.uk/library/documents-w5/rcsg-00.htm
Schafer, B (1998), *Partisan Approaches to Post War American Politics*,
  Chatham House, New Jersey.
Scottish Constitutional Convention Report (1995), *Scotland's Parliament:
  Scotland's Right*.
http://www.cybersurf.co.uk/scotparl/briefing/scc_prop.html
Unit for the Study of Government (1997), *Democratic Participation and
  the Scottish Parliament*, University of Edinburgh, Edinburgh.

13 On the Scottish Road to Sustainability?

KEVIN DUNION

Introduction

The devolution settlement has equipped the Scottish Parliament with powers over a wide ranging portfolio of environmental responsibilities including environmental protection; road building; public transport subsidy; energy efficiency; renewable energy; and indeed for sustainable development. What are the indications that the Parliament is up to the task?

An initial test is whether the new government's proposed policies incorporate sufficient initiatives to indicate that the devolved responsibilities are being taken seriously. The Programme for Government put forward by the Labour/Liberal Democrat coalition suggests that this may be the case. However further tests are whether the driving force is coming from within Scotland or is being externally applied. It will be argued that there are indications that the use of fiscal measures to guide economic activity towards a more sustainable trajectory may remove initiative from the Scottish parliament; and that the overarching need to maintain a conformity of policy north and south of the Border is making for homogeneity notwithstanding the existence of a coalition government in Scotland. That is not to say there are not some signs of a distinctively Scottish position being taken.

The perception of external drivers may encourage a backlash from within the Parliament and more widely, against such policies. However even where powers are available to the Parliament there are indications that there is a lack of enthusiasm for environmental initiatives and there is not yet a political culture which appreciates the scale of the challenge which sustainable development presents, and which has been devolved to Scotland. As a consequence the Executive has been slow to create a sense of dynamic and to put in place the architecture which will generate the momentum equivalent to the task.

Sticking to the Script

Looking back, when the negotiating teams from Labour and the Liberal Democrats sat down together over the weekend of 8/9 May 1999, few political pundits would have expected that the partnership agreement which was thrashed out would have contained so many commitments to tackling environmental needs. After all, the environment had hardly featured in the election campaign, prompting concern amongst the major environmental campaigning organisations in Scotland. They commissioned journalist Rob Edwards to conduct a survey of the four Parliamentary parties in Scotland (excluding, with no sense of prescience, the Green Party and Scottish Socialists). The intention was to try to generate some news coverage of the environment during the election, but also it has to be said to winkle out public commitments which may be useful to refer to once the parties were in position of authority.

The results of that survey are interesting more in the way they were handled by the parties than in the detail. Ten questions were asked including: do you think there is a need for a second Forth road bridge; should the superquarry proposed for Harris be given the go-ahead; what targets would you set for cutting carbon emissions; do you support the plan to establish National Parks in Scotland. The Scottish National Party response was provided by Alex Neil, who held the environmental portfolio. Perfunctory answers were scribbled alongside the original questions and faxed back on the same day. No to the second Forth bridge; no to a superquarry ; targets for carbon emissions 'per Kyoto'. [1]

By contrast it was feared that Labour might never reply. Two weeks after the deadline Jack McConnell provided a 3 page response which caused some disquiet amongst the environmental groups. The question on the Harris superquarry was ducked on the grounds that:

> ...the Report of the Public Inquiry into the proposed superquarry on Harris is outstanding... We will decide our attitude to the final decisions when that Report has been published, giving proper consideration to the recommendations made. [2]

No acknowledgement then of the fact that, electorally, Labour had previously stood on a platform in Harris against the quarry, nor that its keynote UK environmental statement 'In Trust for Tomorrow' had explicitly committed to halting superquarry developments. [3]

The answer on the Forth bridge was however even more disconcerting. The Tories might be expected to back it (their spokesman Ian Stevenson said 'although we rejected the original plan for a second

Forth road bridge at Queensferry we would not rule out a second crossing if a suitable site elsewhere could be found').[4] The others, it was presumed, would surely oppose it and gain easy green credentials as the scheme had no real prospect of going ahead. Yet McConnell replied:

> Scottish New Labour does not have a specific party policy on the issue of a second Forth bridge. We will keep the competing views for and against such a development under review and assess any proposal which may come forward in the future in the context of our policies on transport, the economy, and the financial situation at that time.[5]

Failure to rule out another Forth road bridge allowed the press to speculate that Labour was not firmly wedded to a policy of moving investment away from road and towards public transport instead. McConnell was furious with this interpretation insisting that his answer simply meant that there was no policy because simply the issue was not on the political agenda to cause a policy to be needed. Yet he chose not to couch his reply in the context of Labour's overall transport policies.

McConnell's answer was clearly that of a man who expected to be in Government (if not in the post of Environment Minister). No hostage to future fortune was going to be offered even if at the time of the election there seemed to be no remote possibility of the bridge going ahead.

Given this fastidiousness it is noteworthy then on how many questions McConnell could give full answers. Fuel poverty would be eliminated within two terms of a Scottish Parliament; Labour would set a Scottish target of a 20% reduction in climate change emissions by 2010 in line with UK government undertakings; house energy labelling would be central to achieving objectives; national parks would be established for the Trossachs and Loch Lomond and consideration given to the case for the Cairngorms. These are however simply a repetition of points made in Labour's manifesto - and McConnell did not stray beyond them.

So, although the environment had been given little public prominence in the election debate, commitments to environmental action were peppered through the party manifestos (although interestingly only Labour made any mention of sustainable development).

Accordingly, when the Labour and Liberal Democrat negotiating teams came together the issue was what wording would be settled upon to translate the environmental policies, into priorities in the partnership programme. The Liberal Democrat team included former party chief executive, Andy Myles, who had recently been recruited to head up the Parliamentary Office of the Royal Society for the Protection of Birds. The Labour team had drafted in Sarah Boyack, who had been elected as MSP for Edinburgh Central. She was a Board member of Friends of the Earth

Scotland and had chaired the commission established by the John Wheatley Centre to draw up a programme for environmental action in the wake of devolution. Certainly it will have been useful to have committed individuals in coming to agreement, but the policy commitments which emerged had already been laid out, with relative congruity, in the election manifestos.

The Coalition Programme - Sufficient to the Task?

The chapter on Environment and Transport in the published programme sets out 4 principles and 14 initiatives.[6] The first principle is 'we will integrate the principles of environmentally and socially sustainable development into all government policies.'

This is an important statement. The devolution settlement passed responsibility for sustainable development to the Scottish Parliament. However, unlike the founding legislation for the Welsh Assembly, there was no requirement on the Scottish Parliament to consider sustainable development in drawing up its programme. The wording of the principle is however somewhat unusual. Sustainable development has commonly been held to mean that which not only weighs in the balance the competing demands of the environment, economic development and social policy, but achieves development which meets environmental, social and economic imperatives at the same time. However, this definition seems implicitly to presume that by development we mean mainstream economic development, tempered by environmental and social considerations. It offers no further articulation of what are the principles of environmentally and socially sustainable development; nor to suggest what monitoring or sanctions would be put in place were government policies to breach such principles. We shall return to this later.

The fourteen initiatives were later amplified in the launch (some would say relaunch) of the Programme for Government on 6th September 1999. Called 'Making it work together' the programme for transport and the environment set out nineteen proposals. Ten of these were given timetables and provided some more detail compared to the initiatives set out in May earlier that year. So for example the original initiative to establish national parks for Scotland now read 'By 2001 we will establish the first national parks for Scotland in Loch Lomond and the Trossachs' - consistent with McConnell's pre election commitment. Similarly the initiative to 'promote the use of renewable energies' had developed to 'we will provide locational guidance on renewable energy development by summer 2000.' However, the main additional information was on transport with the promise of a bill in early 2000 to invest in transport and to encourage bus use. It said

> This Bill will allow road user charging and charges on parking at the work place, where it is sensible to do so. We will use the money raised to invest in transport improvements.

- exactly the same wording as used in Labour's manifesto. (The Liberal Democrats manifesto had said they would 'give the power to local authorities to introduce road pricing schemes and tax non residential parking.') Alongside that were some promises such as details of an action plan to develop through ticketing by March 2000 and to establish a framework for a national transport timetable system by the end of 1999 with implementation by the end of 2000. By March 2002 the new programme promised to have moved an additional 15 million lorry miles per year off the roads in Scotland by increasing the freight facilities grant.

On the basis of a never-mind-the-quality, feel-the-width consideration, it could be concluded that the environment holds a higher place in the government programme than suggested by the political discussion prior to the election, but in reality, this is wholly in line with what the manifestos offered.

Despite this, a number of substantial anxieties remain regarding the inclination and the capacity of the Parliament to deliver the scale of radical change to bring about sustainability in Scotland.

The first is whether or not the key figures in the Government beyond the Environment Minister, have any proper appreciation of the scale of the challenge confronting Scotland. This is not the place to retrace international policy developments and debate over the nature of sustainable development. However the Scottish agenda has to contribute to global efforts to constrain northern overconsumption which is now the centrepiece of negotiations at the UN Commission for Sustainable Development; and the OECD. The magnitude of the North - South divergence is captured by a few headline indicators. The OECD countries have a population of c1.1 billion persons, some 19% of the world's population; the world low income group a roughly similar number around 21%. Yet the OECD countries account for almost 83% of world GNP, 81% of world trade; have 92% of private cars, and account for 75% of world energy use; 80% of iron and steel, 81% of paper, and consume over 50% of global grain production. By contrast the low income group has 1.4% of world GNP; 1.0% of trade, and 500 - 800 million of them are chronically malnourished. 100 million are without adequate fuel (Carley and Spapens p42). As the United Nations Environment Programme acknowledges 'the global human ecosystem is threatened by grave imbalances in productivity and in the distribution of goods and services. A significant proportion of humanity still lives in dire poverty, and projected trends are for increasing divergence between those that benefit from economic and technological development and those that do not' (UNEP 1999 pxx). The challenge then is for rich nations to constrain their consumption to within their ecological space, leaving the capacity of poor nations to expand their consumption whilst still living within the ecological limits of resource use and capacity to absorb emissions.

The consequence for Europe (if a target year of 2050 to achieve equity is adopted) is that CO2 emissions would have to be reduced by 77%, and use of non renewable materials such as cement, pig iron and aluminium by 85-90% (Carley and Spapens 1998 p106). For Scotland the scenario is roughly similar (Friends of the Earth Scotland 1996).

One of the essential principles of sustainable development, then, is that for developed nations, economic activity must be constrained within a framework which delinks growth from over consumption of resources and sets targets to bring about an equilibrium where such activity not only respects the limits imposed by the carrying capacity of the Earth, but observes the moral imperative to redistribute the share of such resource use amongst the world's population.

The impact of such a policy trajectory is plain in the international negotiations on climate change emissions. Whatever the inadequacies of the Kyoto and Buenos Aires agreements, there is observably a basic tenet which is that the developed nations must reduce their climate change emissions, effectively delinking economic development from carbon energy use.

The suspicion remains that within the Scottish Executive and its prominent agencies sustainable development has not been approached with the recognition of the radical change involved. In part this may have been because Scottish politicians and civil servants have had little if any role in international dialogues and negotiations, from which UK commitments are derived. This in turn may have encouraged a default position of consistently seeking to avoid disrupting business as usual by doing little more than is required by law, rather than by best practice, and by adopting broad definitions which dilute the concept. Highlands and Islands Enterprise for example was criticised in a report for another Government agency, Scottish Natural Heritage of having a 'particularly idiosyncratic' definition of sustainable development as meaning a sustainable economy, businesses and communities as well as a sustainable use of natural resources.[7] Jim Wallace the Scottish Executive Justice Minister has told the Parliament:

There are a number of definitions of sustainable development. Perhaps the best way I can put it is that sustainable development is about economic growth, social development and environmental protection.

What is not yet evident in Scotland is a recognition that these objectives have to be achieved simultaneously, and also within a certain timescale and with overarching recognition of the ecological limits. Rather there is a suggestion that these objectives have to be borne in mind when making decisions, weighed in the balance but that regularly one may outweigh the others, even if it has deleterious effects.

That is not to say that constructing a sustainable development strategy does not confront us with hard choices and the need for new ways of working.

There is a genuine debate to be had, for example, about how to deal with overconsumption when sections of our society remain marginalised geographically or socially (Boardman et al 1999). Such concern is appropriate but does not excuse us from aiming to construct policies which respect the needs of the poor without permitting or encouraging over consumption by others. Nor should it be the basis of a political culture which seeks to opt Scotland out of measures to meet sustainable development imperatives.

This culture may be generated in response to an externally driven agenda exacerbated by policy developments which sees a shift away from the command and control mechanisms of regulatory provision (which, for example, limit emissions to air and water), in favour of economic levers being used to alter business behaviour. This shifts the agenda away from the Scottish Parliament.

For, notwithstanding that sustainable development is a devolved responsibility, many of the major levers remain with the Westminster government. Indeed one of the key policy developments of the Brown Chancellorship has been the preparedness to consider fiscal mechanisms to bring about environmental improvements: the climate change levy may be followed by an aggregates tax and by a pesticides tax. It remains to be seen whether these will have any differential impact on Scotland. The risk for the Executive is that although the capacity to apply such measures is largely a reserved power, the backlash is being devolved. Conservative and SNP attacks on the Scottish Executive over the fuel duty escalator give an indication of what may be in store. Firstly there was the charge that Scotland had not been taken into account in constructing a policy which works best where there is an alternative in the form of public transport and secondly that Scotland is harder hit from such charges because they are designed to tackle car use in the congested metropolitan areas of England rather than in the sparsely populated parts of Scotland. Scottish Ministers were left to defend a policy which they had little opportunity to influence and where they had no capacity to exercise the fine tuning which would have dealt with genuinely remote populations such as offsetting fuel duty with reductions in Vehicle Excise Duty for certain postcode areas. Politically, Brown defused that particular argument by removing the fuel duty escalator for the whole of the UK, but the stage has been set for future such battles. We can for example expect water charges to rise with the implementation of EU directives uniformly across the whole of the UK and the climate change levy has already caused one of Scotland's major energy consumers, Caledonian Paper Company, to suggest that it would leave Scotland if the levy was implemented as initially proposed.

The Executive appears flat footed and disengaged from this debate. The climate change levy may actually see significant fiscal advantages accruing to Scotland. It is proposed that increased costs of non domestic energy use be offset by reductions in employers National Insurance Contributions. Businesses and activities with high personnel costs and no primary energy demand such as banking, insurance, call centres. local authorities, higher education establishments may find a net gain. Yet there has been little discussion about what should be done with these windfall revenues, nor whether the net gain will inhibit the price signal effect of increased energy costs and thus mean that Scotland is unlikely to see energy efficiency investments which the levy is intended to prompt, beyond the energy intensive sectors of paper or chemicals. What then for the commitments to curbing climate change emissions?

The question remains unanswered as to the degree of Scottish engagement in policy discussions leading up to these fiscal measures. The aggregates tax for instance was postponed for more than two years while the Treasury negotiated with the Quarry Products Association over industry proposals for a voluntary scheme which would deliver outcomes similar to that expected from the tax. Amongst the outcomes deemed desirable was to shift extraction away from sensitive landscapes and in particular national parks. In the negotiations English environmental groups have pushed the Treasury to go further and protect Areas of Outstanding Natural Beauty (AONBs). If this line had been adopted there would be consequences for Scotland since we currently have no National Parks (and no significant quarrying activity takes place in the earmarked designated sites of the Trossachs and Loch Lomond) and secondly AONBs are a strictly English designation. Scotland's highest landscape designation has been the National Scenic Area, and it is precisely in such an area that the Lingerbay superquarry has been proposed. The effect unless account was taken of

Scottish designations, would have been to encourage developers to come north. The question was: had Scottish ministers or staff considered the implications of such proposals?[8] The question is probably not - although the matter became moot when the 2000 Budget abruptly brought to a halt the negotiations by announcing a tax of £1.60 per tonne which would largely apply across the board.

The same question can be asked of mineral policy generally. It has been announced that there will be a review of Mineral Planning Policy in England. This has enormous implications for Scotland as it was Mineral Planning Guidance 6 (MPG6) in England which strategised that coastal superquarries in Scotland (and Norway) would be required to meet the demand for aggregates in the south east of England. Even under the Conservative administration Scottish policy was more cautious. However National Planning Policy Guidance 4 (the Scottish equivalent of MPG6), made provision for up to 4 coastal superquarries to meet projections of significant increases in English and continental demand. Since then demand has actually fallen (Cowell et al 1998) and the only existing superquarry, at Glensanda, continues to operate at well below capacity. Given the changed political and market circumstances it might be expected that there would be an assertion that it is not for the Westminster government to presume what would be the planning policy in Scotland. Yet far from a robust line being taken on the English review the Executive has indicated that it does not intend to alter the non coal elements of the Scottish mineral planning guidance. This apparently leaves it the Coalition policy to make provision for a market demand which has not risen as projected and to require local planning authorities to consider the best sites for coastal superquarries - to which both the Coalition partners were thought to be opposed.

Another apparent example of the Executive not appearing to be engaged in framing policy for which it nominally at least has devolved responsibility affects one of the most high profile environmental issues of recent years: genetically modified organisms. Trials of GM crops have been conducted at sites in Scotland. Consideration of such trials is undertaken by the UK Advisory Committee on Releases to the Environment. However sanction for the trials has to be given by the Scottish Executive, and where they go ahead they are regulated by the Scottish Environment Protection Agency. In 1999 a case was successfully brought against the Department of the Environment, Transport and the Regions for having permitted trials of winter oilseed rape to go ahead without going through the relevant advisory and regulatory processes.[9] The Secretary of State for the Environment at Westminster, Michael Meacher, admitted the government had acted illegally. What remained unclear was the situation in Scotland. It was suggested by the Scottish Rural Affairs Minister that any decision in an

English court did not apply to Scotland even though he admitted that there were four sites which were being used for the trials, and which had been subject to the same variation in permission as had been successfully challenged in England.[10] Whilst the legal systems are separate, politically there is every indication that complete conformity with English policy was being pursued, even to the extent of breaching the law. Again it is open to doubt whether any separate Scottish consideration was given to the trials (even though such matters are a devolved matter) and whether there was any Scottish consideration as to whether the variation in procedure may expose the Scottish Executive to legal challenge.

Prior to devolution and for the period of Conservative government it may be thought that the role of the Scottish Office was to fit Scottish circumstances to UK policies; now however there is a presumption that policies have to fit Scottish circumstances - whether they are developed north or south of the border. So far it would appear that there is room for improvement in engaging with UK and indeed English legislation which has policy implications for Scotland, particularly in areas which are intended to be devolved.

There are of course many instances where a Scottish dimension is being developed. The review of opencast policy in Scotland was generally felt to equip us with a regime which if anything was tougher than the outcomes of the English review, after a period when opencast applications north of the border were on the increase due to what was perceived as a more permissive planning environment. The proposals on freedom of information are held to be more progressive than those proposed by the Home Secretary in Westminster.

Perhaps the most encouraging development however was the decision by the Parliament's Transport and Environment Committee to hold an inquiry into the siting of mobile phone masts. It was seen to be responding to a popular concern which had left the Executive unmoved. Its recommendations - that the masts should require planning permission - that the potential health impacts should be a matter taken into account by planning authorities who should exercise the precautionary principle by siting away from e.g. schools and densely populated areas - is a challenge to the Executive. If the Executive does not respond to the Committee's recommendations, it will be interesting to see whether the Committee will use its powers to initiate legislation. These powers are a characteristic of the Scottish Parliament which marks the committee process out from its Westminster equivalent, and provides even more scope to have distinctively Scottish legislation and disturb the Executive's desire for conformity.

Encouraging though these examples are, they tell us little. Prior to devolution there were plenty of examples of the Scottish Office adopting policy and regulations which differed from England. So whilst the democratic deficit has been addressed it does not unleash a pent up environmental radicalism.

Nor is the enactment of the Coalition's own principles and policy initiatives - welcome though they may be, sufficient to convince that we are embarked upon a period when the Executive is driving towards a sustainable Scotland. For instance the national waste strategy has been criticised for passing the buck to cash strapped local authorities, and similarly the Executive's plans for tackling congestion and climate change emissions from transport seem to rely upon local authorities offering to implement charging schemes. It has been warned to start governing rather than just administering.

A Framework for Sustainable Development

What is missing from Scotland is any architecture which will provide the sense of scale and proportion to what we are attempting when we say that Scotland is committed to the path of sustainable development. The Secretary of State for Scotland's Advisory Group on Sustainable Development, in its final report, concluded that its advice to the new Parliament should not be a wish list of new legislation but rather should highlight the need to get the structure right for delivering - and being seen to deliver- sustainable development. Its recommendations have provoked an interesting response from the Executive.

When the coalition took office there were four key areas where urgent action would appear to be needed

Firstly, setting targets and indicators. We need to know where we are heading and how we can establish measures of progress. For some reason the Scottish Office was reluctant to establish indicators of sustainable development despite considerable evidence of their use elsewhere and pilots in local authorities such as Fife and Strathclyde. The publication of 'A Better Quality of Life' by the DETR has set out what it calls The UK Strategy for Sustainable Development. It appears however that within the Scottish Executive, it is viewed as an English document and that Scotland can set its own targets and indicators. In which case the temptation is to adopt as indicators that which is already measured and to adopt as targets commitments which the Government of the day has already given. This is not sufficient to the task. The challenge is to dare to project what has to be done and to measure progress whilst acknowledging that outcomes may fall short of what is required.

Even where indicators have been adopted, it is going to be essential to judge what will be the impact on sustainability of specific Government policies, and not just through a historical data set. So the second piece of superstructure is some form of audit commission which can independently comment on the likely or actual effects of Government policies in combination or singly, which are worthy of detailed examination because of their capacity to affect sustainability targets. (In the manner which the House of Commons Environment Audit Committee currently does.) It would have been interesting to know how much of the increase in CO_2 emissions from transport was due to the increased number of journeys at peak travel time caused by education policies which permitted parents to chose schools outside of traditional catchment areas. (The City of Edinburgh Council Local Transport Strategy suggests that the number of children being driven to school has increased four fold between 1971-1990, and that one in five urban trips in the morning peak are taking children to school.) It would be better and illuminating if such estimates could have been made at the time of the policy being adopted and if the Government of the day was obliged to address how it was going to compensate for the damage to its sustainability targets.

Indicators and assessments inform policy choices but do not always prescribe them. Two further necessary functions are a forum where key protagonists can debate and wrestle with issues, and a source of informed advice to Government.

The AGSD did have a range of interests in its membership drawn from business, local authorities, academics, quangos and environmental NGOs. However to be frank it then fell between two stools; the representative nature of the membership did not guarantee that up to date informed advice based upon current thinking and action on sustainability was available to the government. And the dynamic of the group was one of wishing to seek consensus and provide constructive input to the Secretary of State which meant that issues were chosen where consensus might be reached and robust partisan engagement on behalf of the sector from which the participants were drawn was discouraged.

A separation of the two functions is required. One requirement is a policy advisory group which is well informed and engaged with current debate and research at a UK European and international level. It should be able to highlight forthcoming issues which may not yet feature on the Governments indicator list. It should anticipate the impact of international and EU agreements on Scotland, considering specifically any distinctively differential impacts for Scotland. It could scrutinise claims of good policy and practice elsewhere to see what lessons can be applicable to Scotland.

Alongside this group however should be capacity for structured engagement to grapple with key issues which affect Scotland particularly where there is evidence of division and doubt over the appropriate way forward. This needs to be more than the traditional consultation process, where written responses are collated and summarised by a civil servant. It has to be more than the occasional meeting under Chatham House rules where the Minister tries to get to the heart of the matter by inviting protagonists to speak out openly, by assuring them that statements are off the record. Rather it needs to be a structured dialogue facilitated by a specialist body with skills and experience to explore the nature of the dialogue and to identify from the outset what groups want from it, to then set ground rules for such an engagement. Appropriate methodology which gives voice to those who are often excluded, and probes entrenched positions should be applied. There would be no requirement for groups to sign up to a common conclusions and reduced scope for fearing co-option by engaging with other vested interests. Such a structure would help to explore contentious and heated issues such as North Sea fishing restrictions; applications of environmental levies; changes in agricultural support. It could also help strategise future policy such as jobs from the environment.

Early in 2000 the Executive moved to respond to these needs by establishing a Ministerial Group for Sustainable Scotland. It consists of the Transport and Environment Minister, the Finance Minister, the Minister of Communities and the Minister for Industry, senior civil servants from Environment, Transport and Planning, Energy, Planning and Development Control, Policy and Sustainable Development and, in a challenging development, two external members from Friends of the Earth Scotland and Shell UK. This construct goes beyond what the AGSD was proposing in that instead of a policy advisory body which like a Royal Commission seeks to influence Government, it actually cohorts politicians, and permits a dynamic of internal and external policy perspectives to be played out in front of them. For instance therefore in response to concerns that the manufacturing and tourism strategies do not embody observable commitments to sustainable development the Group has called for Scottish Enterprise, Scottish Tourist Board, Scottish Homes etc to meet with the Committee to be questioned on how they propose to respond to the Executive's commitment to sustainability and how this will be reflected in their spending plans for 2001. This kind of leverage would not be available to an external advisory group.

The Group has also commissioned research into Scottish sustainability indicators finally breaking the resistance to having yardsticks which will show whether or not the Government in Scotland is living up to its commitments.

The argument for a Roundtable is still being made (although it is thought that the Civic Forum may be able to have a role to play); and the function of an Audit Commission is still to be addressed.

Of course these developments do not guarantee that we are well on the road to sustainability. The test of that will be the outcomes not processes, but at least Scotland is now equipping itself within the Executive and through the more accessible functions of the parliament with the levers which can accelerate progress towards a more sustainable Scotland.

Conclusion

The coalition programme contains environmental commitments which, if not recognised in a run up to an election dominated by tuition fees and arguments over tax and spend, were provided for in the manifestos. Some of those policies such as land reform and National Park legislation are clearly determined by Scottish priorities. The capacity to take distinctive Scottish positions is provided for by the extensive nature of the powers devolved and may be enhanced if Committees not only hold the Executive to account but initiate legislation.

Other policies however are intended to maintain a UK conformity, such as in transport strategies, or in application of European objectives in diverting waste away from landfill or making provision for strategic environmental assessment. This may hold dangers for the Executive if Westminster imperatives cause such strategies to change. There may also be a backlash against externally determined agendas, particularly where these are delivered through reserved powers such as application of fiscal measures.

Such a backlash may be not just against the political origination of the environmental agenda but against the agenda itself. It cannot yet be presumed that simply because there is a decent count of environmental policy initiatives that Scottish institutions have appreciated the scale of the challenge, and the contribution which Scotland has to make, if commitments to sustainable development are to ring true.

Notes

[1] Worldwide Fund for Nature Scotland, Friends of the Earth Scotland, Royal Society for the Protection of Birds Scotland. 'Results of a survey of the four main political parties in Scotland April 1999'.

[2] Ibid.

3   Labour rejects development of any further coastal superquarries. We do not
        believe that carving up the coast of Scotland in order to feed the Department of
        Transport's voracious roads programme is rational or sustainable. In Trust for
        Tomorrow - report of the Labour Party Policy Commission on the Environment
        1994 p35.

4   Worldwide Fund for Nature et al. Op cit.

5   Ibid.

6   Other aspects of the Coalition programme which contribute to sustainable
        development are to be found in other chapters e.g. Freedom of Information; land
        reform; energy conservation and building regulation.

7   'HIE vandalising Cairngorms' Sunday Herald 12 December 1999.

8   In fact the question has been put in the Scottish Parliament: Fergus Ewing
        (Inverness East, Nairn and Lochaber) (SNP): To ask the Scottish Executive
        whether it has made or will make representations to Her Majesty's Government
        regarding the impact in Scotland of the proposed Aggregates Tax (S1W-2790).
        Sarah Boyack: 'The Scottish Executive is in regular contact with H.M.
        Government on a wide range of issues including the Aggregates Tax.

9   R v Secretary of State for the Environment, Transport and the Regions and The
        Minister of Agriculture, Fisheries and Food, ex parte Friends of the Earth Limited,
        CO/3398/99.

10   "Friends of the Earth consider Scots challenge over GM tests" The Scotsman 18
        September 1999.

References

Boardman, B., Bullock, S. and McLaren, D. (1999), *Equity and the environment –
        Guidelines for green and socially just government*, Catalyst/ Friends of the Earth,
        London.

Carley, M. and Spapens, P. (1998), *Sharing the World - Sustainable Living and Global
        Equity in the 21st century*, Earthscan, London.

Cowell, R, Jehlicka, P., Marlow, P, and Owens, S. (1998), *Aggregates, Trade and the
        Environment: European Perspectives* IUCN UK Committee.

Department of the Environment Transport and the Regions (DETR) (1999), *A better quality
        of life - a strategy for sustainable development for the UK*, Cm 4345 The
        Stationary Office, London.

Fife Regional Council (1995), *Sustainability Indicators for Fife*.

Friends of the Earth Scotland (1996), *Towards a Sustainable Scotland*, Edinburgh.

New Scottish Labour (1999), *Building Scotland's Future*.

Organisation for Economic Cooperation and Development (OECD) (1998), *Eco-Efficiency*
        OECD, Paris.

Secretary of State for Scotland's Advisory Group on Sustainable Development (1999)
        Scotland the sustainable - 10 action points for the Scottish Parliament, The
        Scottish Office, Edinburgh.

Scottish Conservative and Unionist Party (1999), *Scotland First*.

Scottish Executive (1999), *An Open Scotland - freedom of information - a consultation*
        The Stationary Office, Edinburgh.

Scottish Green Party (1999), *Caring for Scotland - Manifesto for Scotland's Parliament*.

Scottish Liberal Democrats (1999), *Raising the Standard - Scottish Parliament Manifesto.*

Scottish National Party (1999), *Scotland's Party Manifesto for the Scotland's Parliament 1999 Elections.*

Scottish National Party (1999), *A Penny for Scotland - investing more in Scotland's schools, hospitals, and homes.*

The Scottish Office (1999), *Down to Earth - a Scottish perspective on sustainable development* The Scottish Office, Edinburgh.

Strathclyde Regional Council (1995), *Strathclyde Sustainability Indicators.*

United Nations Environment Programme (UNEP) (1999), *Global Environment Outlook,* Earthscan, London.

14 The Scottish Parliament and Scottish Culture: The Opportunity of the New Era

PAUL HENDERSON SCOTT

The Present and How we Arrived at it

From one point of view, the restoration of the Scottish Parliament is a very small and timid step forward. Its powers are so limited and so restricted by Westminster that it is weaker than the subordinate parliaments of most federal states. For an ancient sovereign nation, which has never completely lost its identity and autonomy, it could be said to be more of an insult than a compliment. Can we not be trusted to run our own broadcasting, to say nothing of taxation and foreign affairs?

On the other hand, although timid, it is a step which changes everything. It is the greatest constitutional change in Scotland since the Treaty of Union which was ratified by the Parliaments of Scotland and England in 1707. The treaty had one essential provision: it abolished the Parliament of Scotland and, in theory, that of England as well. In their place, it created the United Kingdom of Great Britain with one Parliament. The Westminster Parliament, so created, has frequently violated the Treaty of Union, but it has never before abrogated the essential provision. Now that it has done so, many possibilities are open. They include the dissolution of the United Kingdom, a constitutional arrangement which depends on the Treaty. As Donald Dewar said at the formal opening of the restored Scottish Parliament on 1ˢᵗ July 1999, it was 'a new stage on a journey begun long ago and which has no end'.

The expectancy and optimism which attended the opening, the feeling that Scotland now enters a new era, are therefore not altogether misplaced. There is an obvious risk that disillusionment will quickly follow. That would certainly happen if the Scottish administration were to insist on following in the footsteps of Westminster and so refuse to find Scottish solutions to Scottish problems and introduce radical changes. On the whole, I think that this is unlikely. One advantage of having our own Parliament on our own doorstep is that public disapproval can easily make itself felt. As Mrs Howden in Scott's *The Heart of Midlothian* memorably remarked: 'when we had parliament-men o our ain, we could aye peeble them wi stanes when they werena gude bairns'. The Parliament is a public forum, for the discussion and resolution of Scottish issues. That is bound to strengthen Scottish democracy and reinforce our sense of identity.

It may not be the dominant idea in the minds of the politicians or of the drafters of their manifestos, but the movement towards Scottish self-government, which has carried us forward to this point, is closely involved in more ways than one with our national culture. In the first place, the fundamental reason why we want to run our own affairs is because we are conscious of a national character which should be free to develop in accordance with its own nature. We believe that we have our own particular set of attitudes, ideas and values, expressed in our literature, music and art. If that were not so, if we were identical to our neighbours, then there would be no case for a Scottish Parliament.

Secondly, there is a close inter-relationship between cultural and political self-confidence and without that self-confidence, demonstrated convincingly in the Referendum of September 1997, constitutional advance would not have been possible. The poet, Robert Crawford, is quoted in *The Herald* of 4 June 1999: 'There is a confidence abroad among Scottish poets. Imaginatively, devolution happened 10 or 20 years ago'. I should say that the imaginative change began much longer ago than that. In 1895 Patrick Geddes wrote about a Scottish renaissance, long before Hugh MacDiarmid gave a new impetus to it in the 1920s and 30s. How it is that ideas and passions, generated by poets, painters and composers, are conveyed to people who know nothing of their work is mysterious. Perhaps it is a two-way process. At all events, it is clear from the record that public opinion tends to follow, sooner or later, the lead of the writers and poets.

Thirdly, we need self-government to remedy the cultural, social and psychological damage inflicted on us by the Union. This an aspect which requires some explanation because we have all been brought up to believe that the Union was a wise and benevolent act of statesmanship which brought nothing but advantage to Scotland. Robert Silver, who was a distinguished scientist as well as the author of a fine play about Robert Bruce, said once that the Scots were 'the victims of the strongest attempt to brainwash a whole nation ever mounted in history'. It has succeeded in obscuring for generations the reality of our situation.

Usually the brainwashing concentrates on the economic case. In the days of the British Empire, with captive markets and imperial preference, there was no doubt benefit for some branches of the economy. There were also careers in the army and colonial service which compensated for the neglect of Scotland itself. All this has long gone and we now live in a global economy with the free movement of goods and people, not just in the small area of Britain, but in the much larger one of the European Union. A comparison of the present condition of Scotland with small independent countries in north west Europe does not suggest that we have benefited from the imperial or British connection. We have lost millions of our people through emigration. We have also lost most of our industry and ownership and control over most of what is left. We have deplorably low standards of public health and housing and high levels of poverty and long-term unemployment.

In cultural terms, with all that they imply for personal contentment and psychological well-being, the damage has been much more insidious. *A Claim of Right for Scotland* of July 1988, the report of the Constitutional Steering Committee which led to the Constitutional Convention and therefore to the Scottish Parliament, begins with a section on 'the essential facts of Scottish history'. One of the conclusions is: 'the Union has always been, and remains, a threat to the survival of a distinctive culture in Scotland'. Recently, a well-informed English observer, John Tusa, in his book, *Art Matters*, made very much the same point:

> For 250 years after the Union, and culminating in the Thatcher years, Scotland was not only subordinated to England politically but culturally as well...Scottish culture had been virtually Leninised - that is to say reduced to a few token, vestigial symbols such as bagpipes, tartan, shortbread, haggis, whisky, a folksy accent and the baccanalian celebration of Hogmanay. It is undoubtedly the case that many Scots were happy to collaborate in the process ...

But there was a rebellious strain of an authentic Scottish identity which was alienated from Thatcherite London and later contributed to the sudden sweep towards the idea of self-government, and possibly even independence. Of course, both the rebellion and the movement towards self-government had very much deeper roots than Tusa supposes. Even so, the close political association with a larger, and a powerful and self confident, country might have led to cultural assimilation. As Colin Bell said at the end of his series of BBC programmes, *Scotland's Century*, Scotland is a nation which could have disappeared, but which refused to die.

The cultural subordination of Scotland is most obvious in language. We have two indigenous languages, Gaelic and Scots. Both have rich literatures. Both, like all languages, embody the shared experiences of the communities which have used them for centuries, and therefore are the best means of expressing aspects of the national character. These languages still exist and are still vehicles of fine literature, but they have been almost entirely displaced by English in education, the law, government and the media. Tacitus said of the acquisition of Latin by the ancient Britons that they thought that it was part of civilisation, but in fact it was a 'feature of their enslavement'. The same might be said of the English language in Scotland.

Certainly, there are many advantages in the modern world in a knowledge of English. It has become so useful a means of international communication that many countries use it as a second language for this purpose. We should want to do the same; but this could have been acquired without generations of Scottish children being reduced to inarticulacy and a sense of inferiority by the denigration of the mother tongue which they brought to school. In other ways as well our educational system has tended to undermine our self-confidence. It has been Anglo-centric, saying little about the extraordinary range of Scottish achievement, but focusing on English conditions, history and literature, and giving the impression that nothing of consequence ever happens in Scotland. The London domination of broadcasting has tended in the same direction. So has the need to look to London for all important decisions in government and business.

The consequence of all this is the so-called Scottish cringe, a massive inferiority complex. Cairns Craig says in his book, *Out of History*: 'Scottish culture has cowered in the consciousness of its own inadequacy, recognising the achievements of individual Scots simply as proof of the failure of the culture as a whole ... And the consequence of accepting ourselves as parochial has been a profound self-hatred'. Iain Macwhirter in *the Sunday Herald* of 27 June 1999 says that: 'Lack of self-respect and self-government have left us prone to a kind of national infantilism'. Of course by no means all Scots have succumbed to such feelings, but there are signs which suggest that they are sufficiently prevalent to amount to a serious weakness. The restoration of national self-confidence must be one of the first objectives of the Scottish Parliament.

What Can the Parliament do About it?

The American sociologist, Michael Hechter in his book, *Internal Colonialism*, argued that England, as the core of the British state, denigrated the cultures of Scotland, Wales and Ireland to justify its political domination and that it used the 'voluntary assimilation of peripheral elites' for the purpose. This is the process which I have been describing. I do not think that it has happened as a result of a deliberate policy, but as the automatic result of the overwhelming size and wealth of England. As Tusa says, many Scots were happy to collaborate and that has included the school teachers. They have done this with the best of motives because they thought it was the best way to prepare their pupils for careers in the British Empire and state. The teaching profession now recognises the consequences of inadequate attention to Scottish history and culture. The Scottish Consultative Council on the Curriculum has drawn up reports to make recommendations on both of these. The first of them has been published, but the second is so far in suspense, apparently because of a sudden loss of courage. Blue prints therefore already exist for the remedy of the Scottish deficit in Scottish education.

It is not only Scottish history, but all history, which has been given too little attention by the schools. This is unfortunate for two reasons. First of all because we cannot understand the present without some understanding of how we arrived here. Secondly because history, properly taught, is an excellent training in the critical examination of the evidence. In this age of sound-bites and high pressure advertising there is more need than ever for the cultivation of the habits of logical, objective and sceptical thought. George Davie in his important book, *The Democratic Intellect*, stressed the importance of the study of philosophy in the Scottish universities in the past. An intelligent approach to history can serve very much the same purpose. With history, as with literature, music and art, we should begin in Scotland because that has most immediate relevance and then radiate outwards to the international connections.

All political parties and people generally agree that the improvement of education is an urgent necessity. In Scotland this is particularly so because, as I have been arguing, it relates so intimately to self-confidence, and therefore to happiness and social usefulness, as well as to the technological skills which the modern world demands. Where once we led, we have fallen behind international standards, and that gives an additional need for urgency. In addressing these problems, the members of Parliament and the civil servants should not yield to the temptation to suppose that they have a monopoly of wisdom. I hope that they will use the committees to seek advice from both the teachers and the taught and from people with ideas.

The Scottish Parliament is responsible for cultural policy; but broadcasting, the most potent means of cultural expression, is reserved to Westminster. This is absurd. In an essay, *The Backwardness of Scottish Television*, Stuart Hood, the novelist and former Controller of Programmes for BBC Television, said that there were societies 'which have not achieved self-awareness because the dominant instruments of mass culture do not provide a mirror in which time citizens can see themselves truthfully'. This is precisely our situation in Scotland. Not only is the BBC controlled from London, but the structure of commercial television ensures that most of the programmes, even of the Scottish companies, are produced in England. For most of the time, we are made to look at ourselves and the rest of the world through London eyes. This was confusing and demoralising enough when our Government and Parliament were also there. It is now intolerable and must be changed.

Many of the most important cultural institutions, such as the National Library, Galleries and Museums, the Scottish Museums Council and Scottish Screen, are state funded but controlled by their own boards. The Scottish Arts Council is such a body. It is responsible for the distribution of Government funds for the support of the arts in Scotland, a hands-off device which avoids suspicion of political favouritism and also shelters the Government from the criticism of unpopular decisions. It is through these organisations that the State has its most significant, if indirect, influence on the cultural life of the country. It can stimulate artistic and intellectual achievement by adequate funding or curtail it by a reduction.

The Scottish Arts Council is a major factor in the cultural life of Scotland, but it has also been fiercely criticised. In its origin it was a sub-committee of the Arts Council of Great Britain and therefore followed their policies. For years it more or less ignored Scottish indigenous music and dance, but this has changed for the better since the Scottish Council became autonomous and since an extensive consultation in 1992 revealed that Scottish traditional music was very much alive and deserved much more recognition and support. A stand-still on funding in the last years of the Conservative government, which was in effect an annual reduction, put great strain on the Council. That was not their fault, but they have also been criticised for too much secrecy, too much bureaucracy and too much arbitrary power. Most of these points can probably be remedied if the Council becomes answerable to a committee of the Scottish Parliament.

On the other hand, it may well be true that the Arts Council has exercised more power than is appropriate for a single body since it became responsible for the distribution of lottery funding for the arts as well as the direct Government grant. I think that there is a good case for the national companies (Scottish Opera, Scottish Ballet, the RSNO and the Scottish Chamber Orchestra) becoming directly funded by the Cultural Department of the Scottish Executive in the same way as the National Library, Galleries and Museums. At present the national companies require such a large proportion of the Arts Council budget that they distract attention from the multiplicity of other companies which are an essential part of the cultural life of the country. The same risk of suspicion of party political favouritism does not apply to the national companies which clearly have to be founded.

Another advantage would follow from this change. It would make it easier to achieve the National Theatre which has been an aspiration for about a century. The case for it was stated very fairly in *The Charter for the Arts* in Scotland, drafted by Joyce Macmillan and published by the Scottish Arts Council as a statement of 'a shared vision' for the future development of the arts into the next century:

> The case for a national theatre rests on the contention that it is absurd for Scotland, which has little indigenous tradition in ballet and opera, to support major national companies in these areas, while having no national theatre to protect and express our much richer inheritance of Scots drama and theatrical tradition. It is also pointed out that, although Scotland supports a rich network of theatre companies, none of them has a specific remit to perform and develop Scottish repertoire and languages, and commitment to it therefore varies unpredictably as artistic directors come and go ...

There is a need for an institution whose remit it is to preserve, develop and promote the Scottish dramatic repertoire, to encourage Scottish writing for the stage, and to help actors and directors acquire and maintain the language and performance skills necessary for the most effective performance of drama in all forms of Scots and in Gaelic. Such an institution would provide a valuable resource not only for the theatre itself, but for all forms of dramatic and literary culture in Scotland, including film and television.

Why then, when the need is so obvious, has the Scottish Arts Council always resisted it and fobbed off demands with yet another feasibility study? I think it is because they have been reluctant to accept responsibility for an additional national company and because of opposition from the existing theatre companies who were afraid of a competitor for Scottish Arts Council funds. Neither of these would arise if all national companies were funded directly by the Scottish Office. In the experience of most other European countries a national theatre has a vital role as a stimulus of the national literature and of national self-awareness and self-understanding.

There seems already to be a real prospect of the expansion of the resources for film making. If that happens, I think that we need only one other new institution, in addition to the National Theatre. That is a body to encourage cultural exchange with other countries in both directions. Until Scotland acquires its own diplomatic representation abroad, this could be combined with offices to promote trade and tourism and there is in fact a relationship between all of these.

No government or parliament, as far as I am aware, has ever written a play, a symphony or a novel, although they have provided material for them. Their main cultural function is to provide the financial support which in the past was provided by royal or aristocratic patrons. The Scottish Government is going to be under very severe financial restraint. It has denied itself the use of even its very limited power to vary income tax and the grant under the Barnett formula is designed to squeeze more tightly year by year. The poor standards of housing and public health will make strong and legitimate demands. On the other hand, the share which the arts need is only a very small proportion of total Government expenditure. Last year (1998) the Scottish Office spent about 0.5% of the total on museums and galleries and about 0.2% on the performing and visual arts through the Arts Council. There is room for improvement without any substantial change in the overall pattern of expenditure.

In any case, the benefit from expenditure on the arts is out of all proportion to the cost. Nothing does more to enhance the international reputation of the country with direct benefit to the tourist trade, inward investment and even to our exports. We have the advantages of having in Edinburgh the greatest international festival of arts in the world and of having our own traditions in literature, music, painting and architecture which are a valuable and distinctive part of European diversity. We should encourage both the international and the national. We need to experience the best that can be produced in the rest of the world, but we have a special responsibility for our own arts because they speak to us directly and because they cannot be found anywhere else. Above all, we should encourage the arts because they enrich the lives of our people, broaden their understanding and contribute to civilisation.

We should study the practice of other comparable countries in encouraging their cultural and intellectual life. Many of them do far more than we have been able to do so far. There is, for example, in the Irish Republic a tax regime which encourages works of art. Norway has a scheme for the purchase of new Norwegian books for distribution to libraries throughout the country.

Financial restraints are not the only problem. There is another which is much more insidious and damaging. Cultural standards everywhere are under the very real threat of the trans-Atlantic fashion of 'dumbing-down'. There are insistent calls for the rejection of cultural values and the acceptance of any music, writing or entertainment as though it were as good as any other. Anything which is difficult and which requires application and effort is condemned. This approach can be represented as a crusade for equality and the rights of the common people, although it is in fact condescending and insulting to them. The use of the word, elitist, as a term of abuse is a warning that this spirit is abroad. Even the Scottish Arts Council is not free from it. A recent paper, *Scottish Arts in the 21st Century* contains this ominous sentence: 'Why are young people more interested in screen-based arts, rock music, and fashion than in the areas which SAC has traditionally supported?'

At the Congress of International PEN in Warsaw in June 1999 the dramatist, Ronald Harwood, spoke passionately about what he called the virus which was attacking the standards of cultural achievement:

> The virus feeds most voraciously on the demands of the market place, to some extent on lower and lower educational standards, and most unfortunately on those who believe that popularity is preferable to anything else ... Anyone who seeks to defend high standards is immediately branded old-fashioned, reactionary, living in the past, out of touch, boring or even fascist ... In all of the media, as a matter of survival, standards are constantly being lowered in order to attract the greatest numbers. Junk food and junk culture are soul mates.

This is a world-wide challenge, but in Scotland we are well placed to resist. We have a long tradition of respect for education and for the broad philosophical approach of the democratic intellect. Our contribution to the arts and sciences is very remarkable for so small a country. As a result of several centuries of these traditions, and of close interchange with intellectual movements in continental Europe, we led many aspects of enlightened thought in the 18th century. There is no reason why we should not be bold enough to aim at a similar role again. We should be open to beneficial influences from other countries; but we should resist the damaging and destructive. Our habits of logical and sceptical thought should be a barrier against mere fashion. I am not alone in this optimism. Peter Jones, Professor of Philosophy in Edinburgh, in 1989 called on Scotland;

> To proclaim itself as unashamedly intellectual, the thinking nation.

He suggested that we should once again not hesitate to give a lead to the rest of the world in critical intelligence and innovative thought, a new Enlightenment in fact. This idea is now widespread. For example, the City of Edinburgh has drawn up a cultural policy which it calls 'Towards the New Enlightenment.'

Public reaction to the official opening of the Scottish Parliament on 1st July 1999 made it very obvious that expectations are high that it will lead to a radical transformation of Scotland. There is no point in having a Parliament unless it makes real improvements. Education and culture are within its responsibilities. Of course, it is not possible to create a new Enlightenment by passing a few laws. That requires a sustained endeavour by our schools, universities, writers, artists, thinkers and people at large; but Parliament can give a lead and help to create favourable conditions. Scotland for too long has suffered from the absence of a body with the power to mobilise and inspire the popular will. In an article in *The Scotsman* on 30th June 1999 Ian Bell said that the Parliament had no hiding place. Among other things it would have to show that Scotland was 'a country that takes education and intellectual ambition as seriously as it likes to claim'. That is the challenge and the opportunity.

15 Devolution and the Political Economy of Scotland

MIKE DANSON AND KAREN GILMORE

Introduction

In the run up to the elections to the re-established Scottish Parliament, much of the economic debate in Scotland did not progress beyond the level of the simplistic headline. In response to this, six economics professors in Scotland wrote to 'The Herald' (30 April 1999) to complain about this approach to the serious economic analysis of the future development of Scotland. Reference in an accompanying article to 'Economists in despair' captured the feelings of our collective depression over the poverty of the treatment of economic arguments by the media, but especially by the politicians themselves. As this showed an almost complete denial of the rich landscape of literature on the political economy of devolution and federalism, this chapter seeks to return to the economic theory of government and policy intervention to try and provide a framework within which analysis can be undertaken of the implications and potentials offered by the Scottish Parliament.

Economic Restructuring

The background to the debate on the alternative development paths facing the Scottish economy has been the massive restructuring of the last quarter century. Prior to the beginning of the 1980s, Scotland would have been defined quintessentially in terms of steel, coal, shipbuilding, heavy engineering and related traditional industries. The subsequent decline of these sectors was mirrored across the UK with a loss of 46% of manufacturing employment between 1964 and 1994 (Brooks and Weatherston 2000); by 1996, less than one in six of the Scottish workforce were employed in manufacturing. The gap in the economy resulting from the decline of the traditional industries has been filled to some extent, but by very different types of industries and employment. Peat and Boyle

(1999) note that it is a number of years since manufacturing has been Scotland's major employment sector, however the shift to the service sector has been marked even in this last decade. The process of industrial and employment restructuring has been closely associated with the increasing rate of introduction of new technology based on computers and the microchip. This is particularly evident in Scotland, the growth of electronics industries in Silicon Glen initially, and later throughout the country, has led to some 46,000 people being employed in the industry and the creation of £13 billion in manufacturing revenues in 1999 (Scottish Enterprise 1999a). In the wider context Nolan and Welsh (1995) show that, despite this, 50% of UK manufacturing can still be classed as low technology, and even much of the electronics sector in Scotland consists of screwdriver assembly plants. In addition, the service sector provides lower 'value added' exports than does manufacturing, accounting for only a third of the value added of manufacturing exports (Blyton and Turnbull 1994).

Such figures may help to give weight to Nolan and Welsh's (1995) argument that Britain in general emerged from the 1980s as a relatively low-skill, low-productivity economy forced to compete in international markets on the basis of low labour costs. This depressing picture is emphasised by Stark's research (1989) which shows that there has been an increasing inequality in wage distribution within the UK. He states that there has been a trend for employment to shift away from middle-wage industries into either high- or low-wage industries, or, into unemployment. This bi-polarisation has probably continued through the 1990s, while the tendency in Scotland has not been out of line with this general picture.

Economic activity, incomes and employment in Scotland are determined to a large extent by the actions of the private sector and, especially in the globalised economy, by multinational enterprises (MNEs) which contribute 34.7% of total Scottish gross value added. Scotland receives a greater share of inward investment than would be suggested by its relative GDP share and the UK outperforms the rest of Europe in attracting foreign direct investment (Peat and Boyle 1999). Between 1991 and 1997, 308 foreign direct investment projects were established in Scotland (Locate in Scotland) with foreign MNEs accounting for 40.2% of all manufacturing investment. Overseas manufacturing companies, alone, directly employ 80,000 people in Scotland. Opinion is often mixed about the desirability of MNEs in a host country. Peat and Boyle, (1999) point out that gross output per employee is much higher in overseas-owned companies and capital expenditure per employee is three times higher than that of UK companies, largely due to the requirements of oligopolists competing in a global market. The number of jobs and investment in employees is highly desirable but there are disadvantages with regard to developing new enterprises, promoting self-generating growth, and attracting higher function jobs.

Beyond the effects of the intrusion of MNEs into the restructured economy of Scotland, not only have the traditional industries disappeared from the landscape, but also the public ownership which had maintained these sectors and Scotland's economy has been lost through privatisation, deregulation and rationalisation (Lee 1995). The reservation of macro-economic powers at Westminster, WTO and EU regulations, the introduction of the euro and the ongoing moves to integrate the EU economies all restrict even further the abilities of the Scottish Parliament to have any fiscal and monetary autonomy, and so real influence over the economy. The profound effects and implications these changes have had on the body politic and on the economic policy agenda should not be underestimated. Compared with earlier times, the possibilities and instruments available to policymakers have changed significantly, usually to restrict their scope and scale (Danson 1991; Newlands 1999) so that the Scottish Parliament is confronted by a very different set of circumstances and industries with which to forge a strategy for a competitive economy.

Despite these changes and limitations, there are still genuine grounds for optimism that a self-sustaining vibrant economy can be constructed. This can be based on three areas: lessons from economic theory, the experiences of other small nations and federal systems in Europe and North America, and the success Scotland has shown in embracing the challenges past changes have imposed on her. As described below, theory and practice suggest that the fortunes of small economies in the globalised economy of the next millennium will be dependent less on the tax and expenditure policies that dominated the elections to the Scottish Parliament, and far more on the 'invisible factors' in economic development which promote cohesion and dynamism.

The Parliament will have a number of implicit effects on the Scottish economy. These will include stimulation in demand as economic activity is boosted in Scotland with visitors and non-local organisations perceiving the need to have a presence in the country. The attraction of a number of new banking and finance operations to Scotland in 1999-2000 is not unconnected with a perceived need to be close to this new arena of activity and represents but the most visible evidence of such inward investment. The consequent increases in employment and incomes will be enhanced through multiplier effects as hotels, personal services and other sectors benefit further from the cash injections by banks, media companies and so forth. There will be negative effects also, with housing and labour market inflation encouraged by this expansion, especially close to Edinburgh (McQuaid 1999), and displacement of jobs and activity within

Scotland towards the capital. Overall there will be a need to be aware of these centripetal forces and to ensure that the benefits are spread across Scotland, but we would argue that such matters are best dealt with under the planning and inclusion agendas as proposed by others in this book. Further, it has already been noted that the block grant from the UK Parliament and so the budget available to the Scottish Executive will be restricted by the application of the so-called Barnett formula to ensure the long term convergence of identifiable public expenditure in the various parts of the UK to reflect assessed needs. Important as this will be to the management of the Scottish economy and its public services, we will not be focusing on the block grant from Westminster nor on the Barnett formula; these subjects are covered elsewhere in this volume.

The opportunity is being taken in this chapter to reorientate the analysis away from the sterility of debating the superficial effects of varying marginal levels of income taxation and the probable size of Scotland's contribution to the UK Exchequer since 1980, in favour of a more political economy approach. In the next section we identify the rationales for government intervention in the economy. In the third section, the different levels or jurisdictions of intervention are discussed in terms of efficiency and effectiveness, identifying the most appropriate according to policy type and purpose. This is complemented in section four with an examination of the different forms of intervention: traditionally seen as taxation and expenditure, but argued here as including other more potent and 'invisible' types of regeneration and restructuring. The final section attempts to pull these themes together.

Economic Intervention: Legitimacy and Need

As exploring the economic powers and potential of a devolved Scottish Parliament involves a new departure in the UK for analysing the delivery of policy, it is necessary to draw on economic theory and experiences elsewhere to explore the possible impacts of such a body and its related potential. Newlands (1997 and 1999), in referring to the American literature and Musgrave (1959) in particular, starts by examining the legitimacy of government intervention in the economy. He identifies the three uncontested neoclassical macro-economic functions for government as stabilisation, allocation and distribution, to which he has added a fourth: the growth function. Newlands then argues that the focus of debate and appraisal of the economic interventions of the Scottish Parliament should be directed to its role in the processes of sustaining growth and development. Having established the legitimacy for intervention, there follows an economic rationale (King 1984; Peterson 1995) for the stabilisation and distribution functions being reserved for the national or higher, supra-national levels, with the greater openness of more local economies making macro-economic management impossible. However, there are sound reasons for the allocation and growth functions being assigned to sub-central government. Moreover, a sub-national level of government may also benefit those beyond its own boundaries, including the wider 'national' economy. It is important to stress this as, in many countries and communities, devolution especially if only for particular areas, or the development of specific new regional powers, can lead to 'backlash' effects and other reactions which may undermine any universally favourable developments. These analyses based on successful and stable models elsewhere can be enlightening and show how even partial devolution can be applied to the advantage of the country as a whole. This line of argument explicitly links the key issues of taxation, expenditure and economic growth for a sub-national layer of governance, and for the Scottish Parliament in particular.

Level of Intervention

The erosion of regional powers, a general lack of spatial competence, the imposition of top-down policies, and the globalisation of the economy at all levels have been blamed for the failure of regional economic planning to deliver improved growth and regional convergence in the post war period. In a recent renaissance, these have been addressed across and *'within the European context* by a move back towards to enhanced regional autonomy and responsibility' (Roberts 1997, original italics). The European dimension is significant in generating this enhanced role for the region. Roberts *inter alia* has argued that the role of the structural funds, and the gradually evolving partnership model as the preferred delivery mechanisms, have played a critical part not only in establishing regional networks of agencies and authorities, but also in creating the conditions and environment for quasi-regional governance structures. Across the EU, these have undoubtedly been instrumental in persuading some national governments, including the UK government with regard to the English regions, of the benefits of greater devolution.

This approach is echoed by Harvey Armstrong (1997) who, to analyse the economic dimensions of the proposed English regional assemblies, has considered the characteristics of multi-jurisdictional planning frameworks. These are most significant in determining the optimal assignment of functions. He confirms the theoretical case for regional level institutional intervention for the purposes of regional planning and development, but also stresses the need for flexibility to allow for the different interests involved. A role is also determined for the sub-regional level, with local government increasingly stressed within these new forms of regional constitutional governance (Bukve 2000; Fairley 1999; Sinclair 1997).

Cappellin (1997), in examining the nature of federalism and the network approach to national regional policy, views these as a means of facilitating the co-operation and/or competition between different national and regional institutions. His analysis echoes policy developments across many of the territories of the EU by focusing on a new public-private partnership approach to regional development policy and planning within a federal arrangement. In other words, he highlights the need for interaction between different levels of government and territory. While the aim of the partnership approach is to create flexible mechanisms of governance of those public-private relations in order to establish a strategic agenda for change, almost implicitly Cappellin and others argue that it can be achieved through the establishing of networks which stress the transactions between institutions and partners rather than the older prescriptive or centralised 'top down' system.

Roberts, Armstrong, Cappellin *inter alia* therefore highlight the connections between the different areas of governance, the micro-, meso- and macro-levels often by design interacting in ways which make it difficult to separate out the varying forces and forms of influence.

Haynes, Maas, Stough and Riggle (1997) in studying the experience of regionalism within the federal systems of government in Germany, Canada and the United States have demonstrated the problems in establishing and maintaining regional coherence in policy and planning. Their most important lessons for the examination of the powers of the Scottish Parliament are that a devolved or federal system does not automatically imply that the regional dimension is recognised nor that appropriate arrangements will flow out of central or federal institutions to benefit the regions. Rather, different economic structures and emphases of institutions and policies arising from different levels of regional representation and empowerment mean there is a need to consider the needs of areas explicitly; this implies the needs assessment underpinning the Barnett formula should be developed to satisfy not only Scottish requirements but those of all the regions of England also. Experiences of regional economic development within such governance systems throw up some important issues for explicitly planning intervention at the regional level in a unitary state. It is crucial to acknowledge that policymakers are dealing with regional jurisdictions which are in a constant process of dynamic realignment. The devolution settlement should be anything but that, therefore; in economic policy terms flexibility is required if the advantages are to be realised.

Somewhat tangential but still relevant to the consideration of the economic development of Scotland under devolution, is the study of city networks in Germany by Eser. As these represent a form of strategic planning within a federal system, they can be viewed as a form of regional planning mechanism which can secure a strategic perspective on meeting the needs of economic development in localities where economic spillovers exist. In some ways the lessons from this analysis point to the benefits of partnerships within Scotland of similar areas, between urban areas in Scotland and elsewhere in the UK and beyond, and so forth. However, this stresses the extent to which normative issues can be integrated into the networking approach to regional planning and development.

This consideration of the forms and levels of intervention confirms from both centralised and federal systems that there are convincing arguments for a strong degree of devolution of economic development powers to the Scottish Parliament. In calling for such moves, Armstrong sees the bulk of the case in terms of functional allocation decisions. He argues for economic redistribution and financial targeting powers to be reserved for higher levels of government, the UK and EU, with 'national' and European taxation and expenditure used to achieve a better balance of economic development. With a greater emphasis on local self-sustaining regeneration, increased competition for mobile capital, and the accepted benefits of regional development agencies (Halkier and Danson 1997) and sub-national governance structures (Danson, Halkier and Cameron 2000), he demonstrates that economic devolution is necessary. Knowledge, expertise, co-ordination, the promotion of linkages and multiplier effects, innovation and policy economies of scale are all more effectively encouraged and located at the local or regional level. Synergies at the local government or Scottish level can then be transferred to the efficient delivery of policies, with the promotion of growth of the economy being rested with these lower levels of government. The question then is how to use these powers to promote new business start-ups, endogenous growth and initiatives to address social and economic exclusion most advantageously for the Scottish and wider UK economies.

Forms of Intervention

This suggests the Scottish Parliament should be seeking and should expect to be able to influence the allocation of resources productively within

Scotland and to have real positive effects on the levels of growth in the Scottish economy. To achieve these benefits, however, will require positive interventions in the economy (as demonstrated by the fiscal federalist literature) and an increase in Scottish public expenditure, as we shall argue below. To do this will require an increase in taxes in Scotland, given the effective Barnett formula freeze on the block grant coming to Holyrood. To address the need for such finance, McGregor *et al* (1997) have demonstrated that the imposition of higher direct taxes in Scotland could lead to an expansion both in the number of jobs and in output in Scotland, with no adverse effects on the Scottish balance of payments, unemployment or inflation. The assumptions underpinning this analysis are not unduly restrictive and recognise that Scotland has a very open economy, as admitted above. The critical conditions to achieve this range of benefits are not that stringent and would not be alien to the politics of the 1960s and 1970s, nor critically to many other member states of the EU. Simply they require that the Scottish workforce does not seek to regain their immediate real standards of living through higher money wages when faced with an increase in direct taxes. In other words, there needs to be a social contract between the government and the workforce, or more specifically with the trade union movement in Scotland.

Examples of social contract and partnership between government and British trade unions go back further than the 1960s and 70s. Kessler and Bayliss (1998) comment that union co-operation during the Second World War had been readily forthcoming; the unions had ceased to be the opposition and became almost partners at governmental level and within industry. Such co-operation was seen to be necessary to support the war effort and union involvement was also desirable in the aftermath in the bid to build a better Britain without the inequities and poverty of the inter-war period. This co-operation for the wider good is evident in such economic interventions as Britain's first prices and incomes policy in 1948, which was endorsed by the TUC. The unions' role in price and wage stabilisation was also required during the 1970s, indeed the slowing of inflation during 1975/1976 was due to initiatives suggested by the TGWU and subsequently adopted by government. It also significant that many of the problems associated with these experiments in partnership were due to international issues such as the Korean War and OPEC quadrupling the price of oil in 1973, factors which were outwith control of national governments rather than problems with the social contracts themselves. Less evident but even more important, perhaps, are other results of such co-operation. The 1960s and 70s were periods of groundbreaking progress in employment legislation. Improvements such as the 1974 Health and Safety at Work Act, the Employment Protection Act and equal opportunities legislation were due in large measure to the work and effort of the British trade union movement.

Ironically it could be argued that the most successful example of social partnership was partially introduced by the British trade union movement to the benefits of another country. British trade union leaders were instrumental in the construction of the social partnership model of post war Germany. While this model has recently been criticised for its lack of flexibility, the degree of industrial relations consensus created by the model was a large part of Germany's post war economic success. The model of social partnership nurtured by Germany has also become the blueprint for many of the policy measures of the European Union. To the extent that Scotland faces the need for such consensus economic management, the partnership model is necessary and appropriate to Scottish civic society (McCrone 1992).

Given the benefits of financing an economic expansion based on a social contract, an alternative and more efficient way to raise taxes in Scotland can be discussed. The Polish economist Kalecki (1937), a contemporary and sometimes critic of Keynes, argued that taxes on capital or property were a more productive approach to growing an economy than the traditional labour taxes favoured by supporters of demand management. Indeed, more recently it has been proposed (Laramie and Mair 1997) that increases in the Council Tax, or more generally a property based source of finance for the Parliament in Scotland, could be used to finance a beneficial expansion of the economy, a similar conclusion to the McGregor *et al* analysis. Unlike the direct 'tartan tax', it would be more difficult to avoid this tax, it would be less regressive, and the sums raised could be more efficiently collected than through the Inland Revenue, with more of the additional income staying in Scotland. However, there would be questions over the attitude of the Treasury to a potential subsidy to the Scots Exchequer from the resulting greater flows of Council Tax Benefit to Scottish households, and over the regressive nature of such indirect taxes. To address these problems, local authorities could be given powers to recognise the need to protect the poorest from any unfair additional tax burden, or new forms of land value taxation or a Scotland-wide local income tax could be established; each of these could reintroduce a fairer degree of redistribution into the tax system. Again, as with the tartan tax analysis, there would be a balanced budget multiplier effect on the Scottish economy (McGregor *et al* 1997). As both capital and revenue public expenditure should have a lower propensity to import than the affected taxpayers (McNicoll 1992), the increase in Scottish government expenditure would result in an increase in spending on Scottish goods and services, with a subsequent multiplier effect across the Scottish economy.

As significantly, there would also be a expansionary effect on the regional economies of England. In terms of the economic spillover effects, noted by Armstrong (1997) as being significant in benefiting the other regions in the rest of a devolved economy, the transmission of the advantages to the Scottish economy of expansion furth (beyond) to the English regions, and to the south-east especially, should not be underestimated. As often claimed, the UK capital region is the most important trading partner for Scotland, and managed growth in Scotland could benefit the south-east of England with limited inflationary impacts.

As argued by Kellermann and Schmidt (1997) *inter alia* experiences in federal countries such as Germany, the US and elsewhere show clearly that tax-sharing and similar financial arrangements between regions can benefit all parts of the country, and not just the apparent areas of subsidy, special deals, etc. They describe different mechanisms and philosophies for transferring finance and resources between regions within states with devolved administrations. Critically, financial flows in these circumstances are considered to be the costs or the lubricants necessary to maintain the viability of the federal nation state, but they can also promise advantages to all parts of the country. Their analysis suggests that a tax-sharing system can lead to long-term convergence, a more efficient inter-regional allocation of capital with a possibility of higher national aggregate growth. In other words, traditional tax and spend policies are not just right for the country as a whole, but specifically within the devolved UK system, can lead to an expansion of the Scottish economy at no risk to the rest of the UK economy, just the reverse in fact.

Partnership

The economic theories outlined above suggest how the government could promote the sustainable expansion of the Scottish economy and so allow the UK economy to run at higher levels of activity than otherwise, avoiding the need to raise interest rates to dampen down demand in one part of the UK while others are still in recession. However, this would be dependent on a genuine partnership with the trades unions and employees in Scotland to avoid jobs being priced out of markets through attempts to regain real wage levels. In its specific industrial and training programmes, also, the Government increasingly recognises the importance of consensual supply-based strategies to future well-being and competitiveness, with such concepts as the 'know-how economy', 'endogenous growth', 'the knowledge economy', institutional capacity building, clusters and so forth, all promoted as critical to the sustainable development of British business

(DTI 1998; Scottish Enterprise 1999b and c). For both the immediate health and the long term growth of the Scottish economy, therefore, there is a need for co-operation, rather than confrontation, for the inclusion of the trades unions in a social contract. So, economic management has to incorporate wider issues than simple market mechanisms, the political economy of devolution must embrace the idea of partnership between all the stakeholders if it is to deliver all its potential.

One of the central planks of the Labour Party campaign running up to the 1997 General Election was the concept of a stakeholder society. Tony Blair repeatedly mooted the idea of a new kind of politics and a new kind of economics in which the principle was that the route to economic success rested in giving everyone a 'stake' in the enterprise for which they worked (Sopel 1997). The manifesto, on which the Labour Party was to go on to win the 1997 election, noted that the key to business success is partnership between employers, employees and unions and that the Government should welcome such developments. This call for partnership from government echoed earlier exhortations from such bodies as the Involvement and Participation Association (IPA) in 1992 and the TUC in 1994 as to the benefits of partnership for the stakeholders involved. Commentators such as Coupar and Stevens (1998) and Pfeffer (1995) quote such examples as John Lewis Partnership, Boots, Land Rover, the FI group and Toyota as companies who have outperformed rivals or achieved a turnaround in performance through a commitment to partnership. Similarly Fernie and Metcalf (1995) have shown a link between consultation and economic performance. Such 'partnership' companies are distinctive in two ways. First, they demonstrate a greater commitment to employment stability and security than other companies and, second, they allow staff more input into the business at an operational and strategic level.

Despite examples of the efficacy of partnership there is little evidence of any re-emergence of the post-war consensus or the type of tripartite corporate structures which would manifest the concept of social partnership. This may be partly due to the changes which would be needed to facilitate partnership. At the heart of the concept of partnership is the idea that employees' commitment and performance will be increased with greater job security. John Monks (1998) points out the problems of achieving this within the present system of corporate governance in the UK which promotes short-term financial profits at the expense of long-term organic growth. Not only job security but other aspects of social partnership could be more easily achieved by considering not just shareholders but other stakeholders in the enterprise: employees, trade unions, suppliers and the local community. The TUC suggest this could be done via placing greater responsibility on institutional shareholders and creating measures to increase the rights of these other stakeholders. Such innovative measures which would put employment security, social inclusion, research and investment, training and economic development above the priority of profit and shareholder dividends would be a welcome step from a Scottish Parliament.

At the level of the region, conurbation and community similar factors have been identified to explain the relative good or poor performance of different areas. Doeringer et al (1987) for instance have captured the essence of this genre in their volume entitled 'Invisible Factors in Local Economic Development'. Michael Porter's study of 'The Competitive Advantage of Nations' (1990) has been applied at several levels of multi-governance systems to demonstrate that, despite the increasing openness of national and regional markets, in an progressively globalising world with restrictions imposed by customs unions, WTO (the World Trade Organization), etc., there are still areas for fruitful discretion in policy terms for local, regional and sub-national governments and their partners. In its favour, during the Thatcher years especially, Scotland introduced and positively encouraged partnerships and networks as models of economic development strategies so that they are well established and effective in Scotland, and capacity has been built in institutions and individuals to an extent not yet achieved in the rest of Britain (Danson *et al* 1999). With regard to public-private partnerships in macro-economic policy areas specifically, Sheila Dow (1997) has considered the role of the financial sector and monetary policy under a devolved government, and draws upon the distinctiveness and historical significance of Scotland's banking sector to argue that major innovations can be introduced in credit policy to the benefit of both industrial and economic objectives. Perhaps most important in her analysis is the identification of the ability of regional governments to intervene in their economies in imaginative ways which fit their constituencies' needs more neatly than the top-down approaches of the unitary state.

Conclusion

Before devolution, the former Scottish Office had some powers to influence the economy through its growth and allocation functions and institutions. The fiscal federalism literature reveals a significant long run potential and now real capacity to intervene beneficially in the economy. It could be argued (Danson 1999) that Scotland has been effective in delivering training and support for new and indigenous businesses (but see Scottish Parliament, 1999 for more critical comment), and in working in partnership across the range of economic development agencies to attract inward investment and to restructure the economy. However, while much has been done to address the redundancies and effects of deindustrialisation over the last two decades, more is required if Scotland is to compete effectively. In particular given the strategy framework outlined here, the reliance on inward investment can impact adversely on the idea of social partnership (Bean 1994). The ability of MNEs to engage in social dumping, such as the example of the Hoover plant at Cambuslang in 1993, militates against consensus and cohesion. The deregulation of the labour market under Conservative governments, 1979-97, made such a phenomenon possible in the UK. In the enlarging Europe of the early 21st century, the ability of MNEs to transfer production in search of cheaper labour costs may force down wage levels in Scotland and, unless MNEs are subject to regulation, their ability to create jobs will be matched by an equal ability to destroy employment, almost irrespective of government grants and subsidies to encourage location in the first place. Additionally, there is a need to guard against the Scottish Executive relying on MNEs for economic development, particularly in disadvantaged areas and hence becoming even more likely to implement policies and legislation which support the company to the detriment of other stakeholders. It could be argued that this was the case during the 1980s, but within the devolved system the Scottish Parliament has the opportunity to implement measures which maximise the benefits of MNEs while protecting the workers and communities who depend on them. The experiences of the 'Celtic Tiger', Ireland, with its low levels of corporation taxes and other incentives to MNEs are pertinent. The long-term boom has transformed the economy, but failed to address poverty and social divisions and to create a sustainable growth (Tansey 1998). This suggests that the allocative efficiency of the Scottish budget must be improved, but also that the benefits of enhanced competitiveness within partnership could be significant (Sweeney 1998). Skill shortages and human capital development especially are subject to ongoing market failure in Scotland and need the level of intervention that only a Parliament can deliver (Scottish Enterprise 1999b and c). From these and other studies, including the Cubie Inquiry into student finance and the McCrone Inquiry into teachers' salaries, the need to reallocate more resources to the vocational education and training budgets is clear. But there is a strong caveat, the relevance of social capital and so good health, education and other public service investments to the attraction and maintenance of the critical dynamic skill groups in the community and to the building of cohesion and consensus can be critical (Storper 1997, Porter 1990, Tansey 1998). There is a need to expand the Scottish budget, therefore, to improve these services and to raise competitiveness. To afford this without damaging public services and civic society, a tax and spending regime has been advocated here based on a social contract or partnership. This could lead to expansion of the Scottish and wider UK economies, with no adverse effects on employment, inflation or inter-regional prosperity.

References

Armstrong, H. (1997), 'Regional-level jurisdictions and economic regeneration initiatives' in Danson, M., Hill, S. and Lloyd, G. (eds.) *Regional Governance and Economic Development*. European Research in Regional Science 7, Pion, 26-46, London.

Bean, R. (1994), *Comparative Industrial Relations: An Introduction to Cross-national Perspectives*, Routledge, London.

Blyton, P. and Turnbull, P. (1994), *The Dynamics of Employee Relations*, Macmillan London.

Brooks, I. and Weatherston, J. (2000), *The Business Environment, Challenges and Changes*, Pearson Education Ltd, London.

Bukve, O. (2000), 'Towards the end of a Norwegian regional policy model?,' in Danson, M.,Halkier, H. and Cameron, G. (eds.) *Governance, Institutional Change and Regional Development*, Ashgate, forthcoming, Aldershot.

Cappellin, R. (1997), 'Federalism and the network paradigm: guidelines for a new approach in national regional policy,' in Danson, M., Hill, S. and Lloyd, G. (eds.), *Regional Governance and Economic Development*, European Research in Regional Science 7, Pion, 47-67, London.

Coupar, W. and Stevens, B. (1998), 'Towards a new model of industrial partnership,' in Sparrow, P. and Marchington, M. (eds.), *Human Resource Management: The New Agenda*, Financial Times Management, London.

Danson, M. (1991), 'The Scottish Economy: the development of underdevelopment?,' *Planning Outlook*, 34, 89-95.

Danson, M. (1999), Economic development: the Scottish Parliament and the development Agencies, in McCarthy, J. and Newlands, D. (eds.), *Governing Scotland: Problems and Prospects*, Ashgate, 87-102, Aldershot.

Danson, M., Fairley, J., Lloyd, G. and Turok, I. (1999), *The Governance of European Structural Funds: The Experience of the Scottish Regional Partnership*, Paper 10, Scotland Europa, Brussels.

Danson, M., Halkier, H. and Cameron, G. (2000), *Governance, Institutional Change and Regional Development*. Ashgate, forthcoming, Aldershot.

Doeringer, P., Terkla, D. and Topakian, G. (1987), *Invisible Factors in Local Economic Development*, OUP, New York.

Dow, S. (1997), 'Scottish devolution and the financial sector ' in Danson, M., Hill, S. and Lloyd, G. (eds.) *Regional Governance and Economic Development*, European Research in Regional Science 7, Pion, 229-241, London.

DTI (1998), *Our Competitive Future: Building the Knowledge Driven Economy*, HMSO London.

Fairley, J. (1999), 'Economic development: the Scottish Parliament and local government,' in McCarthy, J. and Newlands, D. (eds.) *Governing Scotland: Problems and Prospects*, Ashgate, 103-120, Aldershot.

Fernie, S. and Metcalf, D. (1995), 'Participation, contingent pay, representation and workplace performance: evidence from Great Britain,' *British Journal of Industrial Relations*, 333.

Halkier, H. and Danson, M. (1997), 'Regional development agencies in Western Europe: a survey of key characteristics and trends,' *European Urban and Regional Studies* 4, 3, 241-254.

Haynes, K., Maas, G., Stough, R. and Riggle, J. (1997), 'Regional governance and economic development: lessons from federal states,' in Danson, M., Hill, S. and Lloyd, G. (eds.), *Regional Governance and Economic Development*, European Research in Regional Science 7, Pion, 68-84, London.

Kalecki, M. (1937), 'A theory of commodity, income and capital taxation,' *Economic Journal, 47*, 444-450.

Kellermann, K. and Schmidt, H. (1997), 'Regional growth and convergence in a tax-sharing System,' in Danson, M., Hill, S. and Lloyd, G. (eds.), *Regional Governance and Economic Development*, European Research in Regional Science 7, Pion, 210-228, London.

Kessler, S. and Bayliss, F. (1998), *Contemporary British Industrial Relations*, Macmillan London.

King, D. (1984), *Fiscal Tiers: The Economics of Multi Level Government*, George Allen and Unwin, London.

Laramie, A. and Mair, D. (1997), 'Macroeconomic effects of regional tax differentials: a post-Keynesian analysis,' in Danson, M., Hill, S. and Lloyd, G. (eds.), *Regional Governance and Economic Development*. European Research in Regional Science 7, Pion, 173-186, London.

Lee, C. (1995), *Scotland and the United Kingdom: The Economy and the Union in the Twentieth Century*. Manchester University Press, Manchester.

McCrone, D. (1992), *Understanding Scotland*, Routledge, London.

McGregor, P., Stevens, J., Swales, K. and Yin, Y.P. (1997), 'Some simple macroeconomics of Scottish devolution,' in Danson, M., Hill, S. and Lloyd, G. (eds.), *Regional Governance and Economic Development*, European Research in Regional Science 7, Pion, 187-209, London.

McNicoll, I. (1992), 'A New Approach to Modelling the Scottish Economy', Scottish Enterprise, mimeo.

McQuaid, R. (1999), 'The local economic impact of the Scottish Parliament,' in McCarthy, J. and Newlands, D. (eds.), *Governing Scotland: Problems and Prospects*, Ashgate, 149-166, Aldershot.

Monks, J. (1998), 'Trade unions, enterprise and the future,' in Sparrow, P. and Marchington, M. (eds.), *Human Resource Management: The New Agenda*. Financial Times Management, London.

Newlands, D. (1997), 'The economic powers and potential of a devolved Scottish parliament: lessons from economic theory and European experience,' in Danson, M., Hill, S. and Lloyd, G. (eds.), *Regional Governance and Economic Development*, European Research in Regional Science 7, Pion, 109-127 London.

Newlands, D. (1999), 'The economic impact of the Scottish Parliament: possibilities and Constraints,' in McCarthy, J. and Newlands, D. (eds.), *Governing Scotland: Problems and Prospects*, Ashgate, 11-24, Aldershot.

Nolan, P. and Welsh, J. (1995), 'The structure of the economy and labour market,' in Edwards, P. (ed.), *Industrial Relations: Theory and Practice in Britain*, Blackwell London.

Peat, J. and Boyle, S. (1999), *An Illustrated Guide to the Scottish Economy*, Gerald Duckworth and Company Ltd, London.

Peterson, P. (1995), *The Price of Federalism*, The Brookings Institute, Washington D.C.

Pfeffer, J. (1994), *Competitive Advantage through People: Unleashing the Power of the Workforce*, Harvard Business Press, Boston.

Porter, M. (1990), *The Competitive Advantage of Nations*, Macmillan, London.

Roberts, P. (1997), 'Sustainability and spatial competence: an examination of the evolution, ephemeral nature, and possible future of regional planning in Britain,' in Danson, M., Hill, S. and Lloyd, G. (eds.), *Regional Governance and Economic Development*, European Research in Regional Science, 7, Pion, 7-25, London.

Scottish Enterprise (1999a), *Key Facts about Scotland*, Glasgow.

Scottish Enterprise (1999b), *Network Strategy*, Glasgow.

Scottish Enterprise (1999c), *Know-how*, Glasgow.

Scottish Parliament, (1999), *Inquiry into the Delivery of Local Economic Development Services in Scotland: Interim Conclusions*, Enterprise and Lifelong Learning Committee, Edinburgh.

Sinclair, G. (1997), 'Local government and a Scottish Parliament,' *Scottish Affairs* 19, 14-21.

Sopel, J. (1997), *The BBC News General Election Guide*, Harper Collins, London.

Stark, T. (1989), 'The changing distribution of income under Mrs Thatcher,' in Green, F. (ed.), *The Restructuring of the British Economy*, Harvester Wheatsheaf, London.

Storper, M. (1997), *The Regional World*, Guilford Press, New York.

Sweeney, P. (1998), *The Celtic Tiger*, Oak Tree Press, Dublin.

Tansey, P. (1998), *Ireland at Work*, Oak Tree Press, Dublin.

16 The Politics of Devolution Finance

ARTHUR MIDWINTER

The Financial Framework

Designing a suitable financial system for devolved government has been a recurring problem for constitutional reformers. Conservative opposition to the 1970s model was centred around the absence of the fiscal powers necessary to promote responsibility and accountability. Birch's (1984) review of the Labour plan identified the lack of a serious fiscal power as a potential weakness, as it created the potential for the Scottish parliament to 'transfer the blame to the British national exchequer for not giving Scotland enough money to do better' (p 93). The current model has a little more fiscal autonomy than the block grant arrangements in the Scotland Act of 1979, but not much. The tax varying power is equivalent to 4.4% of the Parliament's income.

Devolution has led to greater political attention being paid to Scottish expenditure, partly by Labour MPs in the north of England concerned at the Scottish advantage, and partly by Conservative MPs reacting to their 'Scottish problem', which they saw as reflecting the Scottish 'dependency culture' (Cooper 1995).

That Scotland has higher expenditure needs than the UK as a whole is generally accepted by the three major UK parties, but the scale of those higher needs has not been determined with any precision. Identifiable public spending has been around 20% higher than the UK average (Parry 1983; Heald 1994) since the 1970s. In part this reflected the bargaining skills of Scottish ministers in an era of interventionist government. Indeed, this success has been suggested as the key motivation behind the Treasury introducing a formulaic approach - to prevent further Scottish gains (Heald 1994).

The Scottish Parliament in the main inherits the expenditure programme of the Secretary of State for Scotland. The Scotland Programme consists of a block element (96%) and a formulaic element. The most recent figures are set out below -

Table 16.1 The Scotland Programme, Cash Plans 1998-9

	£ million
Agriculture, Fisheries and Food	517
The Scottish Block	13,979
Nationalised Industries External Financial Limits	26
Total	14,523

Source: Serving Scotland's Needs 1998

Resources for agriculture, fisheries and food are decided for Great Britain as a whole, whilst the nationalised industries are determined individually. The block and formula level is determined by incremental adjustments to the previous year's budget allocation, with Scotland receiving 10.66% of the increase or decrease agreed for English programmes and 10.06% for English and Welsh programmes, representing Scotland's share of the population. With the expenditure base at some 32% *above* the English per capita level, what this means is that Scotland receives *smaller percentage* increases/decreases than the equivalent English block.

This system has been regarded by successive Secretaries of State as protecting Scotland's spending advantage. In part, this is because 'in-year' budget decisions can be taken on a non-formulaic basis, which reflect other criteria such as budget share or unemployment. Heald's (1994) comprehensive analysis of this aspect suggests it has contributed to the *stability* in the Scottish per capita relative, as has population decline. We shall return to this issue later.

Whilst the total is determined through this process, the outcomes can be reallocated according to Scottish priorities. Often, however, these political priorities are similar to the rest of the UK (Kellas 1984; Keating 1985; Midwinter and McVicar 1998). There is good reason to expect greater divergence under devolution.

The Parliament's funding will be through a continuation of this 'block and formula' system. This will ensure that:

- Scotland will continue to benefit from its appropriate share of UK public expenditure;

- the Scottish Parliament's assigned budget is determined by a method which is objective, transparent and widely accepted;

- the Scottish Parliament has the maximum freedom to determine its own expenditure priorities;

- the Scottish Parliament has a defined and limited power to vary central government taxation in Scotland and alter its overall spending accordingly;

- the UK government can maintain proper control over public expenditure and public borrowing at the UK level;

- there are clear lines of accountability for local government spending and taxation;

- UK taxpayers as a whole will be insulated from the effects of local decisions which add to Exchequer-funded expenditure in Scotland (p 21).

In practice, therefore, the new assigned budget is determined mainly by reference to changes in comparable English or English/Welsh parliaments, through automatic adjustments. The main exception to this is the Agriculture, Fisheries and Food programme (AFF) which is negotiated separately. The element determined by payments under EU schemes will be settled annually on the requirements of the programmes. The discretionary elements will be incorporated into the block. The assigned budget is voted by the UK parliament and provided as a grant, negotiated by the Secretary of State for Scotland. In addition, the tax varying powers would permit increases or decreases in expenditure of around £700 millions.

We would expect these arrangements to operate for at least the first five years of the new Parliament. The Government has conducted a referendum on that basis, and recently reaffirmed them, rejecting a call from the Treasury Committee for a new needs assessment study which 'would help to show whether the Barnett formula remains the appropriate method of allocating annual expenditure increases to the four nations of the Union'. (HC97-8, 341).

There is no political consensus around them, however. The Conservative Party remains opposed to any new tax powers for the Parliament, and may well review the basis of determining devolved expenditure in the future. Similarly, the Scottish National Party is likely to argue for increased tax powers in the years ahead. At the core of this issue is disagreement over Scottish needs and resources.

Scottish Expenditure Needs

The decision to maintain the Barnett formula as the basis for determining Scotland's assigned budget has been criticised on a number of counts. Firstly, it has been attacked for 'overfunding' Scotland, relative to expenditure needs (Major 1997). This argument reflects the conclusion of the Treasury's 1979 needs assessment study, which concluded that Scottish spending was some 7% above its relative needs. Proponents of this argument call for a new needs assessment study. Calls for another needs assessment study also emerge from pro-devolutionists, who would prefer any such review to be conducted by a government favourably disposed to the devolution settlement, and reflects a belief in the capacity of needs assessment methodologies to deliver objective results (McCrone 1999; Scottish Council Foundation 1997).

In practice, the art of needs assessment requires a fine meshing of technical and political judgements (Audit Commission 1993), and needs models are capable of manipulation to deliver politically acceptable results (Keating and Midwinter 1994; Carr-Hill and Sheldon 1992). The Treasury Committee undertook a review of the Barnett formula, and its report was inconclusive. It did not demonstrate that Barnett was a fair or unfair approach to determining public spending shares. Most sophisticated systems utilise regression analysis to determine the weights to be attributed to needs factors, but these are difficult to defend, and irrelevant when comparing needs of only *four* countries (Midwinter 1999). As the Treasury evidence to the Treasury Committee observed:

> There is no scientifically objective way of saying which factors justify which level of expenditure - in the event the question of what is fair is a question for political judgement (CH 1998, p 4).

Political acceptability is at the heart of resource allocation. A government ill-disposed to devolution could quite easily find a pretext for change, and no needs study would prevent it. Even the most advanced systems require interpretation and judgement, given the multicollinearity of most need indicators. There is no technocratic solution to this political problem.

Another criticism is that Barnett will lead to expenditure convergence on a per capita basis, and that this is inappropriate given Scotland's higher expenditure needs. This view holds that the consistent application of the formula will lead to expenditure convergence as the factors which prevented this from 1979 to 1992 have now been dealt with by the Treasury through tighter administrative controls on decisions which 'bypassed' the formula, and the 'recalibration' of the formula to reflect population shares at the 1991 Census (Bell et al 1996; Cuthbert and Cuthbert 1998; Kay 1998). These authors offer no empirical evidence of Barnett resulting in 'a squeeze' on Scottish spending in practice. Indeed, the empirical evidence shows the system continuing to maintain the Scottish expenditure relative in the period 1995-6 to 1998-9. They predict a squeeze in future. That said, it has to be recognised that the changes to the formula - to utilise more relevant and accurate population data - were clearly defensible.

This 'squeeze', however does not refer to Scotland's spending *levels*, only Scotland's *share* of the UK budget, in a context of spending growth. It was this element of Barnett which made it attractive to Scottish politicians in 1979 (Parry 1983). Indeed, the basic problem with this critique is the failure to realise that *convergence is not* a policy objective. Successive governments have acknowledged that Scottish spending needs are higher, but the political judgement has been that they are less than current spending levels infer (Scottish Affairs Committee 1993; Heald 1993). The present government has confirmed this position to the present author in private correspondence.

The Barnett squeeze identified by such critics, then, is simply slower growth, and the Parliament's expenditure will grow, whilst Scotland's share of the UK budget will fall from 7.8% to 7.6%. Over the same period, Scotland's population share will fall by 0.15%, broadly in line. Barnett, however, provides substantial advantages to Scotland in terms of financial management. It means that Scotland's share of expenditure cuts in times of retrenchment will reflect its population *share* (not its budget share) and *that any reduction in the expenditure relative will occur in periods of growth, thus not requiring spending reductions*. This property of Barnett has been overlooked by its critics.

The real uncertainty arising from devolution is over the extent to which Scotland will be able to negotiate shares of non-formulaic increases in expenditure. Scotland was strongly represented in the British Cabinet under administrative devolution. Will it be so in the new era?

The Politics of the Fiscal Deficit

At the heart of the political argument is disagreement over the imbalances in Scotland's needs and resources, leading to conflict over the fiscal position. This has become an annual 'set piece' confrontation between the Scottish National Party and the Government around the publication of the *Government Expenditure and Revenue in Scotland* report, or GERS as it is known. The reports have been consistently criticised by the SNP, as a 'bogus' statistical exercise' which is 'thoroughly discredited'. However, Scotland's leading expert on government accounting argued strongly in favour of such production in his memorandum as specialist adviser to the Scottish Affairs Committee (Heald 1993, p 24).

The GERS methodology contains assumptions, and the data requires interpretation and judgement. It assessed total Scottish expenditure at £31.8 billion in 1996-7, giving a figure of 16% above the UK per capita average. Identifiable expenditure accounted for £24.7 billion; non-identifiable expenditure was £3.1 billion; and other expenditure of £4 billion. Scottish revenues were estimated at £24.7 billion (excluding North Sea oil and privatisation revenues) leaving a deficit of £7.1 billion.

This issue is not of direct relevance to devolution finance *within* the UK, where expenditure is allocated according to need, but is crucial to political arguments over the merits of devolution and independence. The SNP assume that Scotland will inherit a fiscal balance which would move into surplus, and their calculations focused on 'additional revenue and expenditure compared to present tax and spending flows'. Indeed, the SNP *used the GERS methodology* in its pamphlet 'It is Scots Who Pay' (SNP 1997), and this became the basis of their claim that Scotland is in fiscal balance. Ironically, the parliamentary answer extracted by the SNP reveals how *fiscally precarious* the Scottish position is without oil revenues.

Table 16.2 Scotland's Fiscal Deficit since 1986-7
(in oil and privatisation income)

Year	£ billion
1986-7	+3.4
1987-8	+4.7
1988-9	+5.1
1989-90	+3.2
1990-1	+1.6
1991-2	-2.9
1992-3	-7.2
1993-4	-8.5
1994-5	-6.5
1995-6	-4.6

Total Scottish deficit since 1986-7	**£11.7 billion**

Source: SNP (1997) It is Scots who pay

In the 15 years from 1979, Scotland had a fiscal deficit of £85 billion, which is converted in a £27 billion surplus when oil and privatisation income is included. This 'surplus' was a central theme of the SNP 1997 election campaign, but it was mainly due to large oil revenues in the early 1980s. Using the SNPs own methodology, Scotland's deficit since then, including oil and privatisation income, is £11.76 billion. The decline of both oil and privatisation income has caused a rapid erosion of Scotland's fiscal health.

The SNP's own budget assessment has been criticised in the press (SNP 1999; Ashcroft 1999; Hutton 1999). Not surprisingly, the SNP have redirected their attacks to the GERS report itself. Particular criticism was made of the change to accounting definitions made in the most recent report, which was attributed to a political desire by the Government to inflate the deficit. However, Scotland's *share* of identifiable expenditure *actually fell* after the change.

Table 16.3 **Scottish Share of Identifiable Expenditure**

Year	%
1992-3	10.40
1993-4	10.43
1994-5	10.55
1995-6	10.49
1996-7	10.36

Source: HM Treasury, Public Expenditure Statistical Analysis 1998-9

This criticism has little merit. The reality is that this change happened earlier in the year in the PESA Report, on the grounds that:

> ... the coverage of the exercise has been significantly extended this year, in order to improve both the *detail* and the quantity of data collected. Expenditure was disaggregated in finer divisions than in previous exercises ... (PESA 1998-9, p 83).

The most developed critique of the GERS methodology has been produced by Cuthbert and Cuthbert, (1998). Their paper assesses the adequacy of the methodology used and the accuracy of data sources,' which they found to be flawed and inadequate. The technical basis of their criticisms is beyond the concerns of the present paper. But these criticisms *of themselves* do not suggest that the GERS conclusions are wildly wrong, but accurate within the 'broad orders of magnitude'. The degree of imprecision in GERS would have to be significant before the judgements made over the scale of the fiscal deficit became questionable (McKay and Wood 1999).

The calculation of the fiscal deficit, however, has become so politically controversial that its publication leads simply to claim and counter-claim over its validity, and this undermines the report itself, which remains a valuable source of public information. It is defined in the document as the Scottish General Government Borrowing Requirement - the gap between expenditure and revenues in Scotland - even though in practice this concept has no meaning, as only the UK GGBR matters. Indeed, the Scottish GGBR will vary with economic decisions taken by the British Chancellor over the desirable and sustainable level of borrowing. This is why the SNP insist in deducting Scotland's share of the UK deficit from the Scottish deficit. All of this is confusing, perhaps even to the cognoscenti!

There is clear evidence from this review that the fiscal arithmetic works in Scotland's favour, but calling the result a fiscal deficit muddies the water and results in polemical debates abut the accuracy of its measurement. This degree of fiscal redistribution is not a sign of economic weakness or a dependency culture, but a reflection of the higher level of social need, and the higher costs of service provision, which arise from socio-economic geography.

Managing Fiscal Dependency

In my previous work, I have characterised the financial relationship between Holyrood and Westminster as one of 'fiscal dependency' (Midwinter and McVicar 1996). This argues that the dominance of the block grant in devolved finances is such that only minimal fiscal autonomy, and therefore weak political accountability, will be the result. A pattern of political conflict over finance is not to be viewed as a recipe for clarifying accountability in the British political system. Heald, Geaughan and Robb (1998) describe this prospect as one of 'perpetual crisis and instability' (p 10).

There can be no doubt that the political case for devolution rests on the prospect of more efficient and accountable government. Tax powers are seen as a key mechanism for promoting more responsive government, and greater accountability to citizens. Devolution is seen as a means of achieving such objectives, whereby:

> The crucial point is that the accountability and fiscal responsibility arguments lead to the same conclusion. Having to make decisions about

taxation has a wondrous effect in concentrating the minds of elected politicians about just how important particular expenditure really is. Fiscal responsibility is thereby enhanced, provided that access to revenues has been 'matched' to expenditure responsibilities' (Heald and Geaughan 1996, p 172).

Britain has a highly centralised approach to public finance. Devolution, however, assumes the merits of difference, and requires tax powers to make it a practical reality. The financial arrangements provide for tax-varying powers at the margin - i.e. over and above the level of assigned budget determined through the block and formula system. This provides for less fiscal autonomy than a local authority, which raises around 20% of its income, compared with at most 4.4% for the Parliament.

Heald's argument is that it is the *existence* of the tax-varying power which matters, not its scale. This is consistent with economic concerns over marginal costs, and is also seen as desirable by the Institute of Fiscal Studies. Financing marginal expenditure is seen as the crucial factor. What does that do for the bulk of expenditure which will be funded by the Exchequer *but* allocated by the Parliament? Arguments about marginal expenditure applied to local government operate in a context whereby local authorities collect *more than just* their marginal costs, and indeed at one time provided over 40% of their income.

Heald and Geaughan argue that the Parliament should have a tax-power, and ideally:

> if starting from a clean slate, it would indeed be desirable for the combined tax raising contribution of the Scottish Parliament and of Scottish Local authorities to finance a higher proportion of total Scottish public expenditure than will be the case (Heald and Geaughan 1996 p 178).

There is therefore much common ground between us. We are both supportive of the need for tax powers to promote accountability. We would both wish to see greater fiscal powers for the Parliament. Heald, a long standing devolutionist, is prepared to accept a minimal power, whereas my view is that the fiscal arrangements will confuse rather than clarify accountability. A recent review of the financial arrangements concluded:

> ....the financial arrangements for the Parliament are inherent in its status as a subordinate rather than sovereign body. It remains part of, and subject to, the UK expenditure planning and control system, and the tax-varying power gives only limited financial discretion. Depending on its composition and the management of political relations with Westminster

this may become a source of friction and instability. It does raise questions of accountability as the Parliament has very limited independent financial capacity, and a temptation may be to blame Westminster for frustrating Scottish aspirations (Mair and McLeod 1999 p 80).

This view is commonly held. Cornford (1996) has observed that the centralisation of fiscal power in recent years will continue, as the tax-varying power is 'being defended because it is so small and not because it is remotely adequate to ensure responsibility' (p 48).

My view therefore remains that the Parliament lacks the necessary fiscal capacity for meaningful autonomy and accountability. There is, of course, an inherent tension between New Labour's approach to economic management, which includes increasingly earmarking monies for modernisation - and is inherently centralist; and the devolution of power. To use Bulpitt's (1982) concept of high and low politics, it is redrawing the boundaries of high politics to leave only micro-decisions on resource allocation to decentralised government.

The irony is that most economic commentators agree that the economic impact of the tax varying powers will be small. The result will be that:

> ....the autonomy which the devolved assembly would enjoy would be strictly limited. So long as the Treasury is charged with the duty of maintaining tight control over the UK economic policy, it is difficult to envisage any substantial increase in this autonomy, particularly as increasing integration of the European economy is building up pressure for fiscal harmonisation rather than fiscal diversity (Bell 1994, p 87).

It is certainly the case that regional government in Europe operates with limited fiscal powers (Blow et al 1996; Keating 1996), and that economic globalisation is resulting in concern for uniformity over business taxation (Heald and Geaughan 1998). It was the prospect of government devoid of fiscal autonomy that has made me critical of the devolution settlement for its capacity to muddy rather than clarify the waters of accountability. The debates over the Scottish National Party's perfectly legitimate proposal to use a penny tax power, attacked as being economically disastrous, and defended as the means of restoring Scotland's education system to its position as the best in the world - with neither scenario being remotely accurate - demonstrate the problems. Labour ministers who defended the creation of the tax powers now attack them as damaging to business. Exaggeration rather than argument, is the order of the day, with no realistic political assessment of the merits of the tax case. The small sums involved could well be utilised in meeting the squeeze from the continuing imposition of efficiency gains rather than funding new services (Midwinter and McVicar 1998).

Assessment

This paper has reviewed the problems arising from the financial arrangements for Scotland's Parliament. In part, these reflect the difficulties of reaching coherent solutions within asymmetrical devolution (Keating 1999), but also continuing tensions between the centralist and decentralist strands of New Labour, keen to present itself as the guardian of the public purse, yet also devolving power.

From a normative perspective, combining responsibility for policy development and revenue raising is a desirable feature of any system of decentralisation.

> .....it is a basic assumption in all systems of decentralised government that the political institutions invented for the territorial subdivisions of the state must, *if they are to have any political credibility*, have some measure of *independence in the level of revenue they raise and the choice of public goods on which to spend it* (Smith 1985 p 100).

The Scottish system is at the minimalist end of the spectrum, with limited tax powers on the European regionalist model rather than the more autonomous American federalist model. In the election campaign, a recurring theme was the argument over the adequacy of the assigned budget provided by the Labour government, and the merits of levying even one penny on income tax.

The argument that it is tax-powers at the margin that matter is difficult to sustain. The limited fiscal discretion that Parliament is such that, once levied, the scope for manoeuvre is further constrained, confirming the IFS observation that 'too much grant dependence curtails freedom of action' (Blow et al p 10) and in this case, further complicates political relationships. The Government has created a 'parliament responsible for local government, but with less tax raising power than the local government for which it is responsible. That is not a recipe for stability in the relationship' (Sinclair 1997, p 19).

If it is the case that economic globalisation and European harmonisation will prevent the devolved Parliament from acquiring a greater degree of fiscal autonomy, then the prospects for accountability are not good. There is a clear need, therefore, for the Parliament itself to review the spending on public services under its control. It already has scope for increasing its degree of fiscal autonomy through control of the non-domestic rates, which in the current year will raise £1.3 billions in taxation, and which have been subsidised from the Scottish Office budget in recent years. Labour in opposition had planned to return this power to local government, but this has been vehemently opposed by the business community. As this would nearly *treble* local authority tax revenues - giving them much greater fiscal autonomy - it is not difficult to anticipate the new Parliament deciding this is a power it would rather retain, as this would allow it to spend any additional revenues or reduce taxation. The economic arguments made over the adverse impact of business rates have never been convincing (Birdseye and Webb 1984; Fothergill et al 1986) and the underlying concept of a 'level playing field' in terms of a uniform tax rate ignores the fact that the *tax* bill reflects both the rateable value *and* the tax rate. To use the populist analogy, even under a uniform business rate, Celtic Park and Old Trafford would continue to attract different tax bills, and it is *tax bills* which matter, not tax levels.

Labour's financial provisions have maintained the spending advantages and high degree of fiscal redistribution derived from the unitary state, although we expect this to remain an issue of political contention. The Comprehensive Spending Review has provided for modest real growth in spending in the early years, although does so in a context of questionable efficiency assumptions (Midwinter and McVicar 1998). It has provided a minimal degree of fiscal autonomy at the margins of expenditure, but the reality is that the decisions on *spending levels* will remain largely the preserve of Westminster, whilst decisions on the *spending mix* will fall to Holyrood. So the new parliament remains as dependent on Treasury decisions as was the Scottish Office. To put this into context, the revenues which would be raised by levying an additional penny through income tax (£230 millions) is roughly the same as the grant reductions suffered by local government since New Labour came to power (Midwinter 1998).

In short, the Parliament's role in increasing political accountability is limited by the system of fiscal dependency. Some will argue that it is preferable for Scotland to have any form of devolved government, however inadequate. However a parliament without a sufficient degree of fiscal autonomy simply blurs accountability rather than enhancing it.

A more promising assessment can be made over the proposals for the accountability of the executive to Parliament. Parliamentary approval of the Executive's budgetary proposals is a key element of accountable government. The Scottish arrangements for the new system were published in the Public Finance and Accountability (Scotland) Bill in September 1999. It has two concerns. Firstly, to set procedures for *authorising* the use of resources by the Executive; and secondly, to hold to account those whose funding comes from the Scottish Consolidated Fund.

The provisions of the Bill reflect the recommendations of the Financial Issues Advisory Group (FIAG). FIAG developed a number of strategic objectives in the arrangements for approving expenditure, managing audit and scrutinising outputs. These were:

- to ensure probity in the handling of the public funds;

- to provide the information which the Parliament needs to make properly informed and timely decisions, and to judge the probity and value of Executive action;

- to provide the public with understandable, consistent, relevant and timely information; and

- to contain overhead and compliance costs.

Some of the 82 recommendations require legislation. Others will be implemented administratively, through written understandings between Parliament and Executive. A key feature of the Bill is the proposals for accountability. The proposals include:

- standardising arrangements for the audit of accounts of certain public authorities under the control of the Public General for Scotland;

- transferring responsibility for the audit of Health Service bodies from the Accounts Commission to the Auditor General for Scotland; and

- introducing mechanisms to enable the Parliament to control the borrowing of certain public authorities.

The Parliament will introduce the system of Resource Accounting and Budgeting currently being adopted at Westminster. It is argued that RAB has advantages of clarity over existing cost accounting systems, the key difference being accounting for resources when used, rather than when paid for. Cash accounting, for example, fails to show the annual cost of capital assets.

The main audit body will be known as *Audit Scotland*, and will consist of the Auditor General, The Chairman of the Accounts Commission, and three other members. The Accounts Commission will continue to be responsible for local authority audit. Parliamentary control will be effected through the Scottish Commission for Public Audit, and the Audit Committee. Within the Scottish Administration the most senior member of staff will be the principal accountable officer, and will appoint other accountable officers in turn.

The audit arrangements include Value-for-Money powers, and the Auditor General may undertake examinations into the economy, efficiency and effectiveness of financial management by public bodies. VFM studies have become well established in the last decade, and are a form of policy analysis. In theory, VFM audit is expected to concentrate on the three "Es:, but in practice, this may require commentary on existing policy and practice.

In conclusion, Scotland's financial position arose from a combination of social needs and the bargaining skills of Scotland's ministers. The Barnett formula retains the simple formulaic approach which had protected the Scottish budget in years of cutbacks. The area of uncertainty is the political clout Scotland will have in the UK Cabinet in the post-devolution area, particularly where negotiations become necessary to find non-formulaic decisions. Labour's approach is consistent with its political centralist approach to public finance, spending according to need, with redistributive taxation the preserve of the centre. This is a minimalist approach to accountability in which the Parliament has greater autonomy on the spending mix rather than the size of the spending cake. Whilst this arrangement may be acceptable to the current British government, there is no reason to assume it would survive New Labour. The reality is that devolution finance remains unfinished business.

References

Ashcroft, B. (1999), 'Are the Nationalists Being Economical with the Truth', *Scotland on Sunday*, 2 May.

Audit Commission (1993), *Passing the Bucks*, The Audit Commission, London.

Bell, D. (1994), 'How Much Money Would Scotland Have?' *Parliamentary Brief*, November, pp 86-87.

Bell, D. Dow, S, King, D. and Massie, N. (1996), 'Financing Devolution', *Hume Papers on Public Policy*, vol 4, no 2 Spring.

Birch, A. (1984), *Nationalism and National Integration*, Unwin Hyman, London.

Birdseye, P. and Webb, T. (1984), 'Why the Rate Burden on Business in a Cause for Concern', *National Westminster Bank Quarterly Review*, February.

Blow, L. Hall, J. and Smith, S. (1996), Financing Regional Government in Britain, IFS Commentary, no 54, Institute of Fiscal Studies, London.

Bulpitt, J. (1982), *Territory and Power in the United Kingdom*, MUP, Manchester.

Carr-Hill, R. and Sheldon, T. (1992), 'Rationality and the use of formulae in the allocation of resources to health care', *Journal of Public Heath Medicine 14*, pp 117-126.

Cooper, J. (1995), 'The Scottish Problem : English Conservatives and the Union with Scotland in the Thatcher and Major Years,' in J. Lovenduski and J. Stanyer (eds), *Contemporary Political Studies 1995*, Political Studies Association, Belfast.

Cornford, J. (1996), 'Constitutional Reform in the UK' in S. Tindale (Ed), *The State and the Nations*, IPPR, London.

Cuthbert, J. and Cuthbert, M. (1998), 'A Critique of GERS: Government Expenditure and Revenue in Scotland', *Quarterly Economic Commentary*, vol 24, no. 1, pp. 49-58.

Forthergill, S. (1986), 'Rates and Employment', report for the Department of the Environment, London.

Heald, D. Geaughan, N. and Robb, C. (1998), 'Financial Arrangements for Devolution' in H. Elcock and M. Keating, (eds), *Remaking the Union : Devolution and British Politics in the 1990s*, Frank Cass, London.

Heald, D.A. (1993), 'The Scotland Programme 1993-4 to 1995-6; Memorandum to the Specialist Adviser, Report of the Scottish Affairs Committee.

Heald, D.A. (1994), 'Territorial Public Expenditure in the United Kingdom' *Public Administration*, vol. 72, pp. 147-75.

Heald, D.A. and Geaughan, N. (1996), 'Financing a Scottish Parliament' in S. Tindale (ed), *The State and the Nations*, IPPR, London.

HM Treasury (1998), 'Public Expenditure Statistical Analysis', 1998-9, The Stationery Office, London.

Hutton, W. (1999), 'Salmond Plays the Fiddle', *The Observer*, 2 May.

Kay, N. (1998), 'The Scottish Parliament and the Barnett Formula', *Quarterly Economic Commentary*, vol 24, no 1, pp 32-48.

Keating, M. (1985), 'Bureaucracy Devolved', *TES Scotland*, 5 April.

Keating, M. (1996), 'Scotland in the UK: A Dissolving Union?', *Nationalism and Ethnic Politics*, vol 2, no 2, pp 232-237.

Keating, M. (1999), 'What's Wrong with Asymmetrical Government', in H. Elcock and M. Keating (eds), *Remaking the Union*, Frank Cass, London.

Keating, M. and Midwinter, A. (1994), 'The Politics of Central-Local Grants in Britain and France', *Environment and Planning C : Government and Policy*, vol. 12, p. 177-194.

Kellas, J. (1984), *The Scottish Political System*, CUP, Cambridge.

Mair, C. and McLeod, B. (1999), 'Financial Arrangements' in G Hassan (ed), *A Guide to the Scottish Parliament*, The Stationery Office, Edinburgh.

Major, J. (1997), 'It's England's Money' *Daily Telegraph* 13 November.

McCrone, G. (1999), 'Scotland's Public Finances from Goschen to Barnett', *Quarterly Economic Commentary*, vol. 24, no. 2, pp. 30-45.

McKay, D and Wood, P (1999) The Economics of Devolution and Independence (Report prepared for Bill Lawrie White, Edinburgh).

Midwinter, A. (1999), 'The Politics of Needs Assessment: The Treasury Select Committee and the Barnett Formula' *Public Money and Management*, vol 19, No 2, pp 51-54.

Midwinter, A. (1998), Local Government Finance in Scotland : Trends and Developments, report prepared for Unison, Glasgow.

Midwinter, A. and McVicar, M. (1996), 'Uncharted Waters: Problems of Financing Labour's Scottish Parliament', *Public Money and Management* April-June, pp 47-52.

Midwinter, A and McVicar, M (1998), 'The Comprehensive Spending Review : Scotland and the Scottish Office', Paper presented to the Third Way to Spend Conference, University of Hull, November.

Parry, R. (1983), 'Public Expenditure in Scotland' in D. McCrone (ed), *The Scottish Government Yearbook 1983*, Unit for the Study of Government in Scotland, Edinburgh.

Rose, R. (1987), *Understanding the United Kingdom*, Longman London.

Scottish Affairs Committee (1993), *The Governments Expenditure Plans 1993-4 to 1995-6: Minutes of Evidence*, 21 April.

Scottish Council Foundation (1997), *Scotland's Parliament : A Business Guide to Devolution* SCF, Edinburgh.

Scottish National Party (1997), *It is Scots Who Pay* (Report), SNP, Edinburgh.

Scottish National Party (1999), *An Economic Strategy for Independence*, (Report), SNP, Edinburgh.

Scottish Office (1998), *Government Expenditure and Revenue in Scotland 1996-7*, The Scottish Office, Edinburgh.

Sinclair, D. (1997), 'Local Government and a Scottish Parliament', *Scottish Affairs* no. 19, Spring pp 14-21.

Smith, B. (1985), *Decentralisation: The Territorial Dimension of the State*, George Allen and Unwin, London.

17 Setting the Pace: Scotland and the UK Devolution Project

PETER ROBERTS

Introduction

The intentions, structure and contents of this chapter are somewhat different to those of other contributions to this book. First, the intentions of the chapter are to provide a broader UK context for Scottish devolution, and to offer an assessment of the implications and consequences for the rest of the UK of what has already occurred and is now taking place in Scotland. Second, the structure of the chapter reflects these intentions: it begins with a brief recent history of devolution in the UK and then progresses to a discussion of some of the more important issues, trends and challenges. It then concludes with an assessment of some possible alternative future pathways for the UK devolution project. Third, the contents of the chapter reflect the wide range of issues, factors and arguments associated with the devolution debate, this focus is unavoidable if the full impact of Scottish devolution on the rest of the UK is to be investigated.

In one sense, this chapter should have been written as three separate contributions: as an element of the introduction to the book, as a component part of the analysis of the consequences of establishing the Scottish Parliament, and as a projection forward of the lessons that are likely to be generated by the implementation of the Scottish approach. However, it can also be argued that it is better to deal with all three topics in one chapter, thereby allowing for a more rounded assessment to be presented of the origins, evolution and experience of the Scottish 'experiment'. It has long been the case that Scotland has set the pace for the devolution of central government functions in the UK and, perhaps more importantly, it has also led the UK in terms of the theory and practice of integrated territorial management and governance. As will be seen later, this combination of a relatively mature culture of territorial policy and strong national awareness, allied with a long tradition of integrated territorial management and a common sense of national purpose, has resulted in the emergence of a model of government and governance that is currently considerably in advance of the UK norm. Even when judged by reference to its pre-1999 status, Scotland operated a higher and more sophisticated level of administrative and political autonomy than other parts of the UK (Wiehler and Stumm, 1995).

Furthermore, following the recent general revival of interest in the political case for the devolution of government, and reflecting the advances that have been achieved in the related fields of regional planning and territorial management, Scotland has retained its position at the vanguard of constitutional change and territorial autonomy. Compared with the devolution package for Wales, for example, the Scottish settlement is more extensive and confers greater powers upon the Scottish Parliament. Gay (1999) has identified three main differences between Scottish and Welsh devolution as set out in the Government of Wales Act:

- the subject areas devolved to Wales are listed in the Act, rather than the powers reserved to Westminster;
- the National Assembly of Wales can make only delegated legislation rather than primary legislation;
- power over devolved areas is transferred to the Assembly as a whole;
- the Assembly can delegate its functions to committees or to the First Secretary.

The extent of the devolution package offered to Northern Ireland is also more restricted than that now available to Scotland and, at present, even this more limited package is evolving more slowly due to the continuing political negotiations. In England, the only significant step towards the genuine devolution of political power that has been taken to date is the arrangement proposed for Greater London, which was implemented in May 2000. As will be seen later in this chapter, the devolution arrangements for regions elsewhere in England have not yet progressed beyond the establishment of voluntary regional chambers and appointed Regional Development Agencies (RDAs).

So Scotland, as in the past, continues to represent the leading edge of the UK devolution project. As such, it is both the standard bearer for devolved government and provides a 'laboratory' for further experimentation with regard to both the technical-administrative and political aspects of territorial planning, development and management. The term territorial planning, development and management is used herein in preference to the alternatives, such as regional, area or spatial planning (see, for example, Healey, 1997). It is a much more complete term than the alternatives and, in addition, although it does imply reference to a specific defined territory, it does not carry with it connotations of cultural or political status and can therefore be applied equally to Scotland, Wales, Northern Ireland and the English regions.

Finally, by way of introduction, it is important to appreciate that the Scottish experience of devolution is not the only influence on the progress of the devolution project elsewhere in the UK. Other influences, from mainland Europe and elsewhere, are evident in the development of the devolution agenda, and these influences can also be detected in Scotland. Whilst Eurosceptics may claim that this provides evidence of creeping Euro-federalism, in reality it simply represents the more open learning environment that has been created due to the presence of closer and stronger links between the nations of Europe. As Bradbury (1997, 26) has argued 'there is a clear case, then, that suggests that European integration in the early-mid 1990s affected the nature of British regional politics and policy in ways previously not considered'.

A Brief History of Devolution

Although devolution as a feature of British political life has been evident since the late nineteenth century, the home rule or devolution debate has faltered or stalled on many occasions. From the failure of the first Irish Home Rule Bill in 1886, the question of devolution for Scotland, Wales and England has frequently been considered as part of a wider constitutional settlement. Uniform devolution across the nation of Britain was seen as a response to the potential constitutional imbalance that would be created by Irish home rule, but, then as now, this centrally-determined approach paid little or no attention to the question of devolution to the English regions.

In place of genuine constitutional reform and the establishment of directly elected government in Scotland, Wales and the regions of England, central governments have instead concentrated upon matters such as the reform of local government, the more effective regional administration of centrally-determined policy and the occasional granting of policy concessions (Keating and Elcock, 1998). These concerns have, however, been limited to those issues that the central 'Imperial core' (Goldsmith, 1986, 167) has been willing to concede, and generally have not amounted to much more than the decentralisation of specific administrative functions.

This state of affairs led to a situation by the early 1990s in which the UK was alone among the larger member states of the European Union in retaining a high degree of central government dominance and control over regional affairs. This can be demonstrated by reference to the evidence provided by Wiehler and Stumm (1995), which is also reflected in Keating's (1998a) assessment, who assigned the arrangements for sub-national government in each member state of the EU to one or more of four categories, with most of the UK assigned to Group 4:

- Group 1 - regions with wide-ranging powers (elected parliament, right to levy taxes, budgetary powers, etc) eg German Länder;
- Group 2 - regions with advanced powers (elected parliament, limited right to levy taxes and limited budgetary powers) eg Spanish autonomous communities;
- Group 3 - regions with limited powers (elected parliament, limited budgetary powers, substantial financial transfers from central government) eg French regions;
- Group 4 - regions with no powers (no elected parliament, all financial resources transferred by central government) eg English counties and Northern Ireland.

The continued concentration of political power, together with the relative inflexibility of public policy that results from such a centralised approach, was seen by the researchers to hinder sub-national areas from responding 'adequately to the new requirements arising from deeper European integration' (Wiehler and Stumm, 1995, 249).

However, anticipating such an outcome from the conservative constitutional processes at work within the UK, a more radical review of the challenges of government and governance associated with the various territories of the UK was developed during the early 1980s. The Alternative Regional Strategy (ARS) (Parliamentary Spokesman's Working Group, 1982) combined devolution with a bottom-up approach to territorial planning, development and management (Mawson, 1997). This document can be considered as the starting point for the current devolution project. The ARS placed the creation of elected governments for Scotland, Wales, Northern Ireland and the regions of England at the top of the reform agenda. The argument used was simple and direct: there was a need to establish powerful executive bodies at sub-national level with substantial budgets and areas of responsibility, these should be accountable to elected sub-national governments. Under such a scenario, the English Regional Development Agencies (RDAs) would be answerable to an elected regional assembly, rather than operating as non-departmental public bodies.

The general principles and certain key aspects of policy that had been developed by the ARS with regard to Scotland, Wales and Northern Ireland, were adopted as Labour Party policy. However, the proposal for the creation of English regional government was not included in the 1983 manifesto and was only mentioned as a commitment to consultation in the 1987 manifesto (Mawson, 1997). What drove the Labour Party to re-examine its stance and to re-visit the ideas contained in the ARS was, at least in part, the growing pressure for constitutional change that became evident in Scotland and Wales during the late 1980s. As a consequence, in 1991 the Labour Party issued a consultation paper on Devolution and Democracy (Labour Party, 1991) and later included a formal commitment to devolution in the 1992 manifesto (John and Whitehead, 1997). Further moves towards the refinement of policy included the establishment of a working party headed by Jack Straw to examine political devolution to the English regions (Labour Party, 1995) and the establishment of a Regional Policy Commission, chaired by Bruce Millan, to consider the creation of regional economic development agencies. The latter body, which reported in 1996, argued the case for the establishment of regional development agencies 'separate from the (elected) regional chambers, but responsible to the chambers and acting as their executive arm in the field of economic development' (Regional Policy Commission, 1996, 33).

At the 1997 election the Labour Party promised less to the English regions than might have been expected, especially given the recommendations made in consultation papers and the findings of the Regional Policy Commission. The commitments to establish a Scottish Parliament, a Welsh Assembly and appropriate arrangements for Northern Ireland (in the event another assembly) were at the fore in the 1997 manifesto, along with the proposal to create a Greater London Authority and RDAs in the English regions. However, when compared with the alternative Conservative prospectus - further administrative regionalism and a promise to defend the Union at all costs (Keating and Elcock, 1998) - the Labour Party programme appeared set to deliver the most radical constitutional reform seen in the UK in the twentieth century. In the event, at least as far as the English regions are concerned, the reality is somewhat less than might have been expected.

Following the election of a Labour government in May 1997, the devolution programme soon gathered pace. In September 1997 referendums were held in Scotland and Wales, and although the level of support for the Welsh proposal was lower than expected, the processes leading to the passing of the Scotland Act 1998 and the Government of Wales Act 1998 were initiated. The Northern Ireland Act 1998 followed, whilst the legislation establishing the Greater London Authority was delayed until 1999. Another early piece of legislation was the Regional Development Agencies Act 1998, which contained provisions related to the establishment of appointed RDAs and the formation of voluntary regional chambers. It should be noted here that the London RDA will ultimately be subordinated to the London Executive (the Mayor and the Assembly) and, in order to avoid pre-empting the decisions of the Executive, the formation of the London RDA has been delayed until the election. In the view of one observer, the 'template to watch is London' (Whitehead, 1999, 104).

So what now exists in the UK is a variable geometry of territorial planning, development, management and governance, what Keating (1998b) calls asymmetrical government. Employing Wiehler and Stumm's typology, Scotland is now operating at an advanced level of devolution, akin to that associated with Group 1 arrangements, Wales and Northern Ireland could be classified at the upper edge of Group 3, whilst the London model just about matches the characteristics associated with Group 3. The arrangements in the eight English regions outwith Greater London still lack any real political powers - no elected parliament exists - and depend fully upon central government to provide financial resources. However, as will be discussed later in this chapter, at least a basis for further development has been established in the English regions. The English arrangements have been variously described as 'a part-work' that does not 'represent a real constitutional settlement' (Benneworth, 1999, 4), 'part of a process' within which the RDAs should be considered to be 'a staging post and not a resting place' (Harman, 1998, 194) and as an opportunity to 'demonstrate the wisdom of making further progress towards effective and democratic regionalisation' (Whitehead, 1999, 102).

It comes as no surprise to discover that the progress of the component parts of the UK towards the achievement of integrated strategic territorial planning, development, management, and governance, has varied. The Scottish case for devolution has always been stronger than that of the English regions, given that Scotland has long-established boundaries, a different legal system, a distinct culture and has been administered autonomously by the Scottish Office (John and Whitehead, 1997). In other words, the Scottish starting point in the devolution race was much more advanced, the political arguments were more compelling, and the case for devolution was more straightforward than was the situation elsewhere. Wales and Northern Ireland also possessed a number of the defining characteristics evident in Scotland, and this situation has been reflected in their subsequent treatment. Many of the English regions, by comparison, were relatively weak in terms of both identity and political power. In part this situation is a consequence of decades (or some would argue, centuries) of centralisation, however, it also reflects the style of regional administration imposed upon the English regions (Roberts and Lloyd, 1999). As ever, the imposition of a regime of administration that is inappropriate and ill-tuned to the needs of a territory, is likely to discourage the emergence of a clear sense of territorial identity. This is certainly the case in some of the English regions. However, as will be seen later, the attempted suppression of English regional identity and politics under the 'centralising impetus of the Thatcher government' (Keating and Elcock, 1998, 6) failed. Rather than reinforcing the power of the centre, the inappropriateness and negative impact of many aspects of Thatcherite policy, especially in the less prosperous areas of the UK, gave rise to a new stronger territorial politics.

A further factor which has worked against devolution in England, is the structure and organisation of the English elements of UK central government. Unlike the situation which existed prior to 1997 in Scotland, Wales and Northern Ireland, the English central departments of state were, and still are, organised on a functional rather than a territorial basis (Roberts, 1997a). This contrasts with the situation in many other European countries, such as France and Italy, where decentralisation of central government to territorial prefectures has been the norm (John and Whitehead, 1997). One important consequence has been that each functional department has determined its own regional boundaries, and this has created territorial inconsistency and the dislocation of central government functions from local and regional requirements.

Key Issues, Trends and Challenges

Having outlined certain of the main points of the complex and confused history of devolution in the UK during the past twenty years, and having indicated the major outcomes and consequences of this history, this section of the chapter identifies and discusses a number of the more important elements evident in the debate on devolution. Particular emphasis is placed on, first, the previous experience of Scottish decentralisation and devolution and the lessons for the English regions, second, the form and nature of the Scottish settlement and some issues of relevance for England and, third, the characteristics of the devolved Scottish model and the evidence of emerging best practice in territorial governance.

However, before examining these issues in detail, it is important to consider the characteristics of the various alternative models of decentralisation and devolved government. In the same way that political scientists distinguish between unitary, federal and confederal political systems at a national level, it is both possible and helpful to identify the various combinations of political and organisational structure that operate at a regional or sub-national level. Although Wannop (1995) and others have identified a wide range of possible alternative models of regional planning, administration and governance, and despite the inherent dangers associated with attempting to reduce complexity to a limited number of categories, it would appear that four basic models of organisation can be derived from previous and current experience. In one sense these models reflect Wiehler and Stumm's (1995) characterisation of sub-national government, whilst at another level, the alternatives reflect the various arrangements of the key elements associated with the design of regional government that have been outlined by Stoker et al (1996) and by Keating (1998a). In addition, by defining these four basic models, and identifying the pathways between the models, it is possible to establish a basis for discussing the possible future evolution of the UK devolution project.

The traditional attitude displayed by many central governments with regard to the requirements of regions or other sub-national areas has been to impose a system of 'regional administration' that operates through a 'neo-colonial' model (Goldsmith, 1986). In some cases this 'regional administration' approach has evolved through the establishment of either 'partial regional governance' - a situation in which although the various regional 'actors' (local government, the private sector, voluntary and community organisations and special interest groups) collaborate in order to plan and manage a region, ultimate political power is retained at the centre and/or at local level - or a system of directly elected sub-national or 'regional government' in which an elected regional government dominates without encouraging the wider participation of the full range of regional 'actors'. In other cases, the evolution of a more 'complete regional governance' can be observed; in this situation directly elected sub-national government encourages the full participation of all 'actors' (see Figure 17.1). Governance in the latter case has been defined as both encompassing the

formal machinery of government

and the informal alliances and networks

through which business groups, environmental groups, neighbourhood groups and amenity societies interlink with formal government and in that way manage aspects of the collective affairs or public realm (Healey, 1997, 8).

What can be seen here is the creation of a direct link between the duties and powers of the formal nation state and civil society in general (Kooiman, 1993).

Figure 17.1 Alternative Paths to Complete Regional Governance

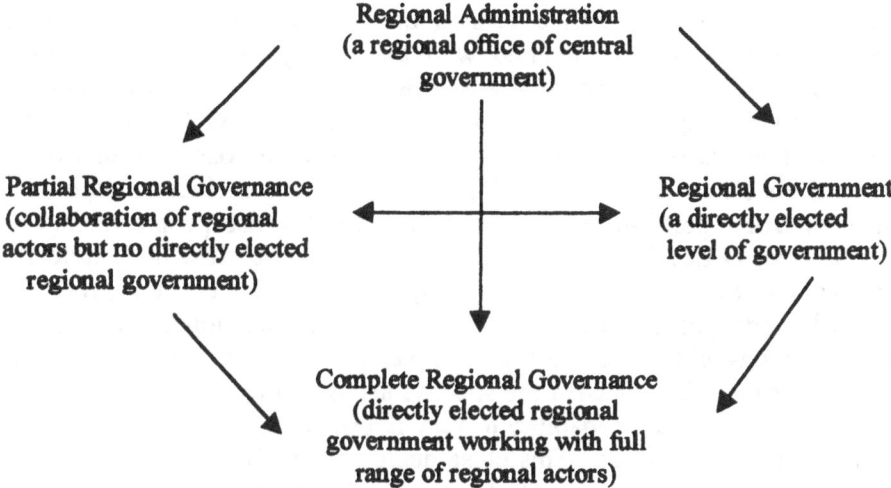

The model (or models) presented above may prove to be helpful in either plotting the course for the future of devolution in the English regions, or measuring the progress of the new arrangements that have been implemented in Scotland, Wales and Northern Ireland. An important feature of the model is that it accepts the availability and possibility of establishing different organisational and constitutional arrangements in the various regions. There is no single tidy formula for the reorganisation of territorial administration or governance in the UK and, as Keating argues (1998b,213) the asymmetrical solution is 'more of a problem in theory than in practice'. Indeed, as is the case with regard to the current RDA 'experiment' in England, a 'one-size-fits-all' approach may fail to take hold in some regions, while in other regions the solution is too restrictive (Benneworth, 1999).

Having established the broad options and possible pathways for the future of devolution in the English regions, the remainder of this section now considers the three issues introduced earlier. The responses to these issues can be organised within the context of the general model.

The first issue to be considered relates to the Scottish experience prior to 1997; a time when Scotland was subject to 'regional administration,' with only a degree of partial regional governance in evidence. In Wiehler and Stumm's (1995) assessment, Scotland, despite not having an elected parliament, could be considered to be a Group 2 territory. Although the Scottish Office was subject to criticism from some quarters as a 'neo-colonial' agent of Westminster government (Kellas, 1991), and despite the suggestion that the arrangement whereby a Secretary of State of Scotland could 'rule' Scotland, irrespective of the representation of the various political parties, was both undemocratic and undesirable, the merits of the Scottish Office version of 'regional administration' were readily apparent when compared with the arrangements in the English regions. The major merit of the Scottish model, in which the Scottish Office was accompanied by other Scotland-wide or regional administrative and development organisations such as Scottish Enterprise, was that most sectoral functions of UK central government were included within the Scottish Office portfolio, to the extent that the 'Scottish people looked to the Scottish Office for answers even when the office had no statutorily defined responsibility' (Mitchell, 1998, 81). In addition, the pre-1997 Scottish model possessed a considerable degree of discretion in the exercise of its responsibilities, an ability to develop and implement policy and policy guidance that matched the specific needs and requirements of all or part of Scotland, and a capability to vary financial allocations in support of specific policy objectives. This close degree of 'fit' between the needs of Scotland and the operation of the pre-1997 public policy regime was evident in a number of policy fields, including the management of the European Union Structural Funds regional programmes (Roberts and Hart, 1996).

Learning from the Scottish experience and applying the lessons to the English regions, the main items that should be considered from the 'regional administration' era include:

- the desirability of extending the portfolio of sectoral issues managed at regional level in order to include all relevant functions - this would involve the expansion of the present Government Offices for the Regions (GORs) model and would require the adoption of common regional boundaries across all central government departments and for all functions;
- the need to reformulate the relationship between the GORs and the plethora of regional and English national agencies and organisations who are involved in the design of policy and the delivery of services at a regional and sub-regional level - this may, in the short-term, include a reconsideration of the relationship between a GOR, a RDA and a regional chamber in order to determine common objectives and procedures for the management of the enlarged portfolio;
- the importance of establishing a stronger sense of regional identity for the GOR and other agencies - Mawson and Spencer (1997, 82) have observed that much remains to be done in 'developing the skills of

(GOR) civil servants in networking partnership development and other forms of organisational capacity building';

- the essential requirement that GORs develop an existence that is independent of the headquarters departments in London - at present a suspicion persists among many regional actors that the GORs remain as 'Whitehall watchdogs' rather than operating as full regional partners (Roberts, 1999);
- finally, it would appear that a pre-requisite for effective 'regional administration' is the presence of a single territorial department of state, however, the reorganisation that would be involved in creating such departments would unnecessarily delay other more radical reforms of the arrangements for the governance of the English regions and cannot be recommended.

A second area of attention with regard to the future of the English regions relates to the nature and form of the Scottish settlement. Given that Scotland moved to its present position from a more advanced situation that that which currently exists in the English regions, it is more difficult to transfer this element of Scottish experience to England. However, a number of general principles are evident and these are worthy of consideration. First there is the question of financial competence, and based on the evidence available from both the general literature and Scottish experience, it would appear that an important consideration in any settlement is the extent of tax-raising (varying) and financial allocation powers (Newlands,1999). A second consideration relates to the old chesnut of *ultra vires*. As was stated earlier in this chapter, the Scottish settlement is different in form and content to the arrangements for Wales and Northern Ireland, chiefly due to the use of the reserve powers approach. By listing the specific areas of activity that are to be devolved, rather than offering general powers except for these issues which are reserved, a devolved government is given less scope and capacity to provide for strategic territorial management. The third and final aspect that can be identified from the Scottish settlement, is the relative ease, despite the earlier political pain and disruption, of introducing devolved government in a territory which has a single tier of local government. This is an issue that has been the subject of considerable controversy in England, with the current Home Secretary, Jack Straw, arguing that you 'cannot establish regional assemblies as well as having shire counties and districts underneath them' (Straw, 1995 quoted in Mawson, 1997). It is somewhat ironic, with the benefit of hindsight, that the previous Conservative administration removed one of the major obstacles to Scottish and Welsh devolution by abolishing the regional and county councils.

The final element of the Scottish case is concerned with the present and the future. One issue is immediately apparent, that is, that Scottish devolution has been accompanied by a new emphasis on bottom-up policy development, partnerships and networks within Scotland. This includes a stated desire to encourage and support innovation in the design of policy solutions and in the development of measures and instruments that are appropriate to meeting the challenges encountered in particular regions and localities. A second theme, which is apparent in a number of recent initiatives, is the greater emphasis now placed on the co-ordination of individual policy measures, especially within a specified territory (COSLA, 1999). From the latter point it would appear that the Scottish experience is in accord with the pattern observable elsewhere: devolution from the nation state to a sub-national level encourages further devolution of powers and functions - this can be interpreted as subsidiarity in action (Stoker et al, 1996).

Although there are many other aspects and themes associated with Scottish devolution that can be used to inform or guide the future of English regional governance, the above examples appear to represent the key issues relevant to the central concern of this chapter. There is, however, one additional observation that should be made at this point. This is, that irrespective of the starting point, the process of devolution, once it has started, would appear to generate its own forward momentum. This can be seen in the evolution of the French regional arrangements (Roberts, 1997b) and, can already be observed in Scotland (Mitchell, 1998). This process of 'rolling devolution' provides support for the arguments presented at the beginning of this section (and illustrated in Figure 17.1) regarding the availability of alternative pathways towards 'complete regional governance'. However, it would be unwise to assume that all the English regions will follow the same route; as both Stoker et al (1996) and Keating (1998b) have noted, there are many other possible roads for both the UK government system and for the English regions.

Possible Future Pathways for English Regional Government

Much of the foregoing discussion and analysis reflects the presence of a substantial body of evidence related to the merits and limitations associated with the alternative pathways to 'complete regional governance' which were presented in Figure 17.1. Although attention so far has been focused on the lessons and organisational models provided by the Scottish experience, this final section also reflects the evidence from other recent cases of devolution elsewhere in the European Union.

Returning to the alternative pathways model, it would appear that the most likely next stage in the development of the current 'regional administration' approach to the governance of the English regions (with the exception of London), is the establishment of 'partial regional governance'. Indeed, it can be claimed that the first step towards the establishment of such an arrangement has already been taken using the powers contained in the Regional Development Agencies Act 1998 that allow the Secretary of State for the Environment, Transport and the Regions to recognise and endorse the formation of voluntary regional chambers. Such chambers, which include not only local authority elected members, but also representatives of the wider regional community (Local Government Association, 1999), have been established in all eight English regions outwith London; different (directly elected) arrangements await implementation in London. Among other reasons for the formation of regional chambers, Murphy and Caborn (1995) have noted the difficulties that have been generated in the past by a fragmented approach to regional administration.

Given that many of the pre-conditions now exist for the establishment of 'partial regional governance', the main question that remains unanswered is: how will further progress be made and how long will it take to establish appropriate regional arrangements? The answer to the first part of the question is relatively straightforward: the chambers are now in place and in some regions, such as the West Midlands, and the North East, there is considerable evidence to suggest the presence of a significant level of joint working between the various key institutions (especially the Chamber, the RDA and the GOR) coupled with a genuine desire to move rapidly to the next stage of evolution. Elsewhere, and especially in those regions that are geographically large and fragmented, or which lack any meaningful tradition of genuine regional partnership working, the pace of progress is likely to be slower, but better this than a rushed arrangement which runs the risk of failure because it cannot take root.

Further devolution, including the creation of directly elected 'regional government' and/or the eventual emergence of 'complete regional governance', which is the product of a mature and open relationship between an elected regional assembly and the wider regional community, is, with the exception of the establishment of London government, unlikely to be on offer in the short-term. Whilst this may frustrate the ambitions and appear to put aside the justifiable demands of these regions that can already demonstrate that they satisfy the criteria for the establishment of directly-elected regional government, the reality is that it is unlikely that Parliamentary time will be allocated prior to the next general election for the passage of the necessary legislation. Despite this, the Labour Party claims that it is still committed to the 1997 Manifesto pledge to introduce regional government: 'we will introduce legislation to allow the people, region by region, to decide in a referendum whether they want directly-elected regional government' (H.M.Government, 1999, 42). Some senior Labour politicians, mostly outwith the Cabinet, continue to argue in favour of introducing directly-elected regional government sooner rather than later, if for no other reason than the need to prevent confusion and conflict between the various regional initiatives and organisations that now exist (Harman, 1998).

So what has now emerged in the UK is a curious patchwork of territorial governance arrangements, varying from a basic or modified version of 'regional administration', to emerging 'complete regional governance' in Scotland. For observers of constitutional change, a strong sense of déjà vu now exists: do you remember the debate on the future of the European Union during the late 1980s? In the late 1990s the same choices appear to be on offer - two-speed regional governance, regional governance a la carte, inner and outer regions, etc; this reflects Keating's asymmetrical approach. One thing is certain, that is, that Scotland is the pacemaker and the test case, although in England considerable attention will also be paid to the performance of the London arrangements. This implies that the experience of devolution in Scotland will both be used as evidence by central government to justify or block further reforms in Wales, Northern Ireland and the English regions, and will be seen by the advocates of change as a source of 'best practice' and as a justification for further devolution. What would be little short of tragic, would be for the cautious steps that have now been taken towards devolution in the English regions to be prematurely judged a failure and abandoned. This is what has occurred in the past due to a collective failure of nerve in central government, the perceived difficulties of implementing 'devolution all round' (Wood, 1998, 8) or the eradication of emergent strong regional administration - such as the Regional Economic Planning Councils of the 1960s and 70s - by a regime that is ideologically opposed to decentralisation (Keating and Elcock, 1998). It is interesting, however, to note that the Thatcher government helped to germinate the seeds of devolution through its scant regard for constitutional conventions and the abolition of regional local government in the English metropolitan areas and Scotland.

Despite the fact that regionalism in England has been variously described as the 'dog that never barked' (Harvie, 1991), and as the likely cause of the 'end of Britain' (Dalyell, 1997), the reality is that devolution has now taken root. If evidence is required to support the case for English regional governance, it is available (see Table 17.1). Public support is also reflected in the views expressed by those involved in attempting to make the best of the present unsatisfactory arrangements. John McCormack, a Northumberland councillor, has put the case for English regional government directly: 'They've taken ten steps forward, we've only taken three or four. We don't want Scotland to slow down, we want to catch up' (McCormack, 1999). This is why the Scottish experience is so important to the United Kingdom at large.

Table 17.1 Identification With An Area
When asked which two or three of these, if any, would you say you most identify with, the answers were as follows (percentage).

	Eng./Scot./Wales	England	Scotland	Wales
This local community	41	42	39	32
This region	50	49	62	50
England/Scotland/Wales	45	41	72	81
Britain	40	43	18	27
Europe	16	17	11	16
Commonwealth	9	10	5	3
Global Community	8	9	5	2
Don't know	2	2	1	0

Source: Mori survey for the economist

References

Benneworth, P. (1999), 'Buying them like shoes: the limitations of the 'one-size-fits-all' approach to the establishment of RDAs in the English regions', Paper presented at the 39th Annual Congress of the Regional Science Association, Dublin, August.

Bradbury, J. (1997), Introduction, in J. Bradbury and J. Mawson (eds), *British Regionalism And Devolution*, Jessica Kingsley, London.

Convention of Scottish Local Authorities (1999), *Best Practice in Community Planning*, COSLA, Edinburgh.

Dalyell, T. (1977), *Devolution: The End of Britain?*, Jonathan Cape, London.

Gay, O. (1999), *Devolution and Concordats,* House of Commons Library, London.

Goldsmith, M. (1986), 'Managing the periphery in a period of fiscal stress', in M. Goldsmith (ed), *New Research in Central-Local Relations*, Gower, Aldershot.

Harman, J. (1998), 'Regional Development Agencies: Not the Final Frontier', *Local Economy,* 13, 194-197.

Harvie, C. (1991), 'English regionalism: the dog that never barked', in B. Crick (ed), *National Identities: The Constitution of the United Kingdom*, Blackwell, London.

Healey, P. (1997), 'The Revival of Spatial Planning in Europe', in P Healey, A Khakee, A Motte and B. Needham (eds) *Making Strategic Spatial Plans*, UCL Press, London.

Her Majesty's Government (1999), *The Government's Annual Report 1998/99*, The Stationary Office, London.

John, P. and Whitehead, A. (1997), The renaissance of English regionalism in the 1990s, *Policy and Politics,* 25, 7-17.

Keating, M. (1998a), *The New Regionalism in Western Europe,* Edward Elgar, Cheltenham.

Keating, M. (1998b), 'What's Wrong with Asymmetrical Government?', *Regional and FederalStudies*, 8, 195-218.

Keating, M. and Elcock, H. (1998), 'Introduction: Devolution and the UK State', *Regional And Federal Studies*, 8, 1-9.

Kellas, J. G. (1991), 'The Scottish and Welsh Offices as Territorial Managers', *Regional Politics and Policy*, 1, 87-100.

Kooiman, J. (1993), *Modern Governance: New Government - Society Interactions*, Sage, London.

Labour Party (1991), *Devolution and Democracy*, Labour Party, London.

Labour Party (1995), *A Choice for England: A Consultation Paper on Labour's Plans for English Regional Government,* Labour Party, London.

Local Government Association (1999), *Regional Chambers: State of Play*, Local Government Association, London.

Mawson, J. (1997), 'The English Regional Debate: Towards Regional Governance or Government?', in J. Bradbury and J. Mawson (eds) *British Regionalism and Devolution*, Jessica Kingsley, London.

Mawson, J. and Spencer, K. (1997), 'The Government Offices for the English Regions: Towards Regional Governance', *Policy and Politics*, 25, 71-84.

McCormack, J. (1997), Quoted in 'England's Upper Lip Fails to Tremble Before Effects of Devolution', *Financial Times*, 07.01.99, 10.

Mitchell, J. (1998), 'What Could a Scottish Parliament Do?' *Regional and Federal Studies,* 8, 68-83.

Murphy, P. and Caborn, R. (1995), *Regional Government for England - an Economic Imperative*, Sheffield Hallam University, Sheffield.

Newlands, D. (1999), 'The Economic Impact of the Scottish Parliament: Possibilities and Constraints', in J. McCarthy and D. Newlands (eds) *Governing Scotland*, Ashgate, Aldershot.

Parliamentary Spokesman's Working Group (1982), *Alternative Regional Strategy: A Framework for Discussion*, Labour Party, London.

Regional Policy Commission (1996), *Renewing the Regions*, PAVIC Publications, Sheffield.

Roberts, P. (1997a), 'Sustainability and Spatial Competence', in M. Danson, M. G. Lloyd, And S. Hill (eds), *Regional Governance and Economic Development*, Pion, London.

Roberts, P. (1997b), 'Whitehall et la Désert Anglais: Managing and Representing the UK Regions in Europe', in J. Bradbury and J. Mawson (eds), *British Regionalism and Devolution*, Jessica Kingsley, London.

Roberts, P. (1999), Developing a Vision, *RDA News*, 2, 14-16.

Roberts, P. and Hart, T. (1996) *Regional Strategy and Partnership in European*

Programmes, Joseph Rowntree Foundation, York.

Roberts, P. and Lloyd, M. G. (1999), 'Institutional Aspects of Regional Planning, Management and Development: Lessons from the English Experience', *Environment and Planning B*, 26, 517-531.

Stoker, G. Hogwood B. and Bullman U. (1996), *Regionalism,* Local Government Management Board, London.

The Economist (1999), *Undoing Britain: A Survey of Britain*, special supplement to The Economist, 353, 8144.

Wannop, U. A. (1995), *The Regional Imperative*, Jessica Kingsley, London.

Whitehead, A. (1999) 'From Regional Development to Regional Devolution', in J. Dungey and I. Newman (eds), *The New Regional Agenda*, Local Government Information Unit, London.

Wiehler, F. and Stumm, T. (1995), 'The Powers of Regional and Local Authorities and Their Role in the European Union', *European Planning Studies*, 3, 227-250.

Wood, E. (1998), *Regional Government in England*, House of Commons Library, London.

Index